'With grit and imagination, Danielle Clode goes on the trail of Jeanne Barret, the first woman to circumnavigate the globe. By chasing down archival leads around the world, and delving into her own maritime past, Clode conjures a spellbinding tale of gender, empire, natural history – and the lure of the ocean. There is poetry and insight on every page.'

YVES REES

'Seamlessly weaving together memoir, history and science, *In Search of the Woman Who Sailed the World* takes the life of its remarkable subject as the starting point for a fascinating and deeply affecting exploration of voyaging, women's lives, and the stories we tell and the stories we don't.'

JAMES BRADLEY

'Biologist, historian, writer, Clode once again demonstrates the connectedness of everything – animals, land, people, plants, sea, sky – at a time when, more than ever, we should be acutely aware of it. Exquisite nature writing brings us close to everyday natural beauty and dramas that we ignore at our peril.'

GAY LYNCH

'A joy to read, simple yet elegant, it whispers in your ear like the sea murmuring from within a shell. With impeccable research, and an ambitious melding of memoir, biography and environmental advocacy, Danielle Clode skilfully reconstructs the life of Jeanne Barret, who emerges from the shadows to her rightful position as "more than a footnote to other men's lives".'

KRISTIN WEIDENBACH

'Danielle Clode unties the knots of myth and weaves a fascinating story of discovery; Jeanne Barret is one of history's most enigmatic explorers.'

NICK BRODIE

'A refreshingly open account that invites the reader on a journey of understanding through the author's own explorations as Clode cleverly reveals the life story of a curious, hard-working, adventuring and independent woman. Clode brings a scientific rigour and a celebration of natural history to the biography of this important woman.'

STEPHANIE PARKYN

Danielle Clode is an award-winning author, biologist and research fellow at Flinders University. She grew up on a boat, worked in zoos, museums and universities and spent many years as a freelance researcher, editor, writer and teacher. Her books include *Voyages to the South Seas*, which won the 2007 Victorian Premier's Literary Award for Nonfiction, and, most recently, *The Wasp and the Orchid*, which was shortlisted for the 2019 National Biography Award.

Also by Danielle Clode

The Wasp and the Orchid
From Dinosaurs to Diprotodons: Australia's Amazing Fossils
Prehistoric Marine Life of Australia's Inland Sea
A Future in Flames
Prehistoric Giants: The Megafauna of Australia
Voyages to the South Seas: In Search of Terres Australes
As if for a Thousand Years
Continent of Curiosities
Killers in Eden

In Search of The Woman Who Sailed the World

Danielle Clode

PICADOR
Pan Macmillan Australia

First published 2020 in Picador by Pan Macmillan Australia Pty Ltd
1 Market Street, Sydney, New South Wales, Australia, 2000

NATIONAL LIBRARY OF AUSTRALIA

A catalogue record for this book is available from the National Library of Australia

Typeset in 12/15 pt Adobe Garamond Regular by Post Pre-press Group
Printed by IVE

Excerpts on pages 105, 117, 132, 205–6, 217, 218–9, 263 and 285 courtesy of Actes Sud (Imprimerie Nationale), 1977.

Picture credit: Map of Jeanne Barret's voyage based on Jacques Nicolas Bellin, 'Carte reduite du globe terrestre', 1764, courtesy of National Library of Australia, MAP RM 2183. Shell images throughout from Antoine-Joseph Dezallier d'Argenville, *La Conchyliologie*. Third Edition Vol. 3, Paris: Guillaume de Bure, 1780, courtesy of Biodiversity Heritage Library.

The author and the publisher have made every effort to contact copyright holders for material used in this book. Any person or organisation that may have been overlooked should contact the publisher.

MIX
Paper from responsible sources
FSC www.fsc.org FSC® C018183

The paper in this book is FSC certified. FSC promotes environmentally responsible, socially beneficial and economically viable management of the world's forests.

Government of South Australia
Department of the Premier and Cabinet

The author gratefully acknowledges research and scholarship funding received from Arts SA and the Department of Premier and Cabinet.

For the seafarers, voyagers and adventurers –
and all those who come by boat

CONTENTS

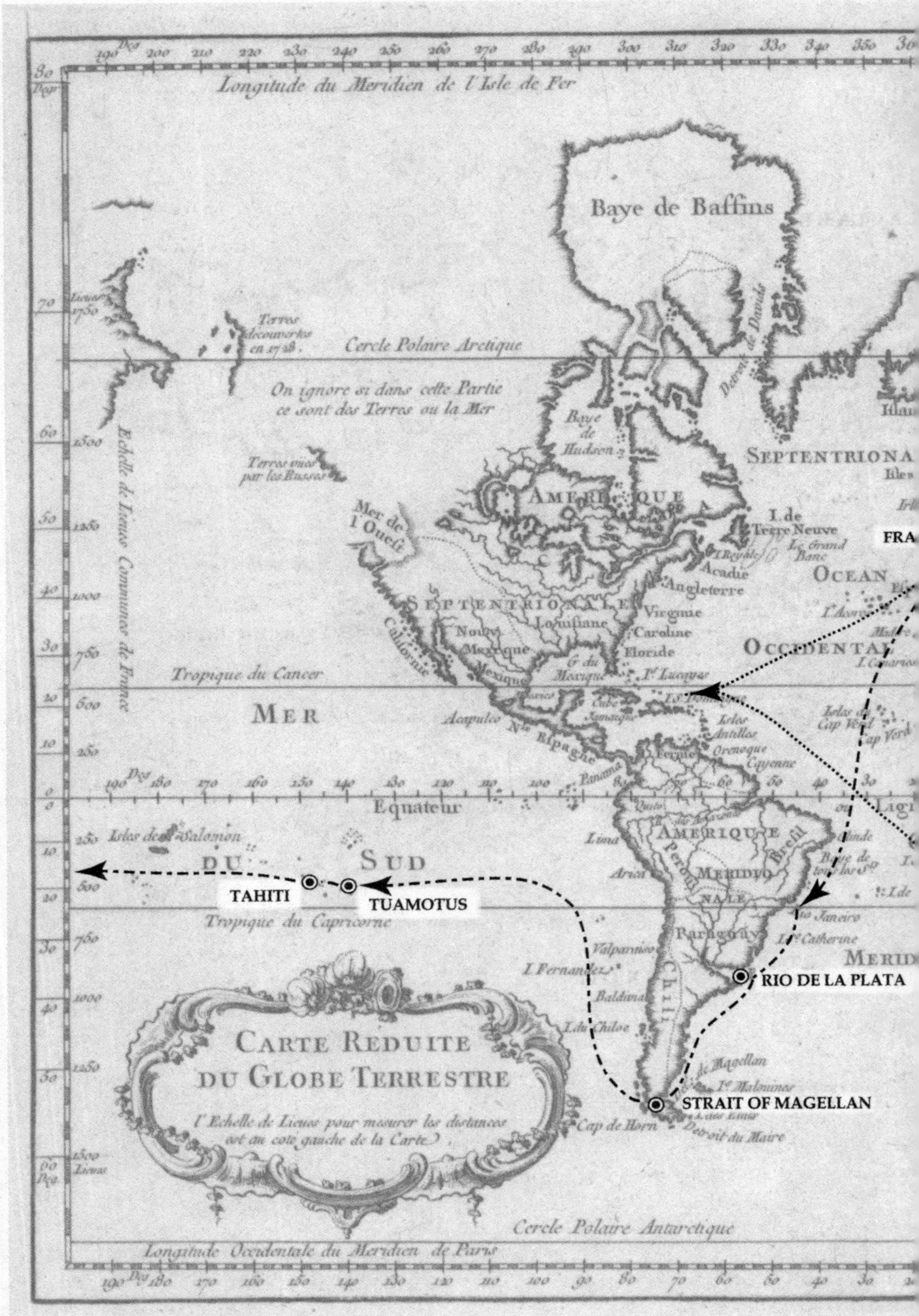
Longitude du Meridien de l'Isle de Fer
Baye de Baffins
Cercle Polaire Arctique
Terres découvertes en 1728.
On ignore si dans cette Partie ce sont des Terres ou la Mer
Terres vues par les Russes
Echelle de Lieues Communes de France
Mer de l'Ouest
Baye de Hudson
Amerique
Septentrionale
Septentriona
I. de Terre Neuve
Le Grand Banc
Acadie
Angleterre
Virginie
Caroline
Floride
Louisiane
Californie
Nouv. Mexique
Mexique
G. du Mexique
Ocean
Occidental
Tropique du Cancer
Mer
Acapulco
Cube
Jamaique
Isles Antilles
Oreneque
Cayenne
Panama
Equateur
Isles de Salomon
Du
Sud
Amerique
Meridionale
Perou
Bresil
Lima
Arica
Paraguay
Chili
Valparaiso
I. Fernandes
Baldivia
I. du Chiloe
Tropique du Capricorne
Rio Janeiro
I. Ste Catherine
Cap de Horn
Detroit du Maire
Carte Reduite du Globe Terrestre
l'Echelle de Lieues pour mesurer les distances est au coté gauche de la Carte.
Cercle Polaire Antarctique
Longitude Occidentale du Meridien de Paris
FRA
TAHITI
TUAMOTUS
RIO DE LA PLATA
STRAIT OF MAGELLAN

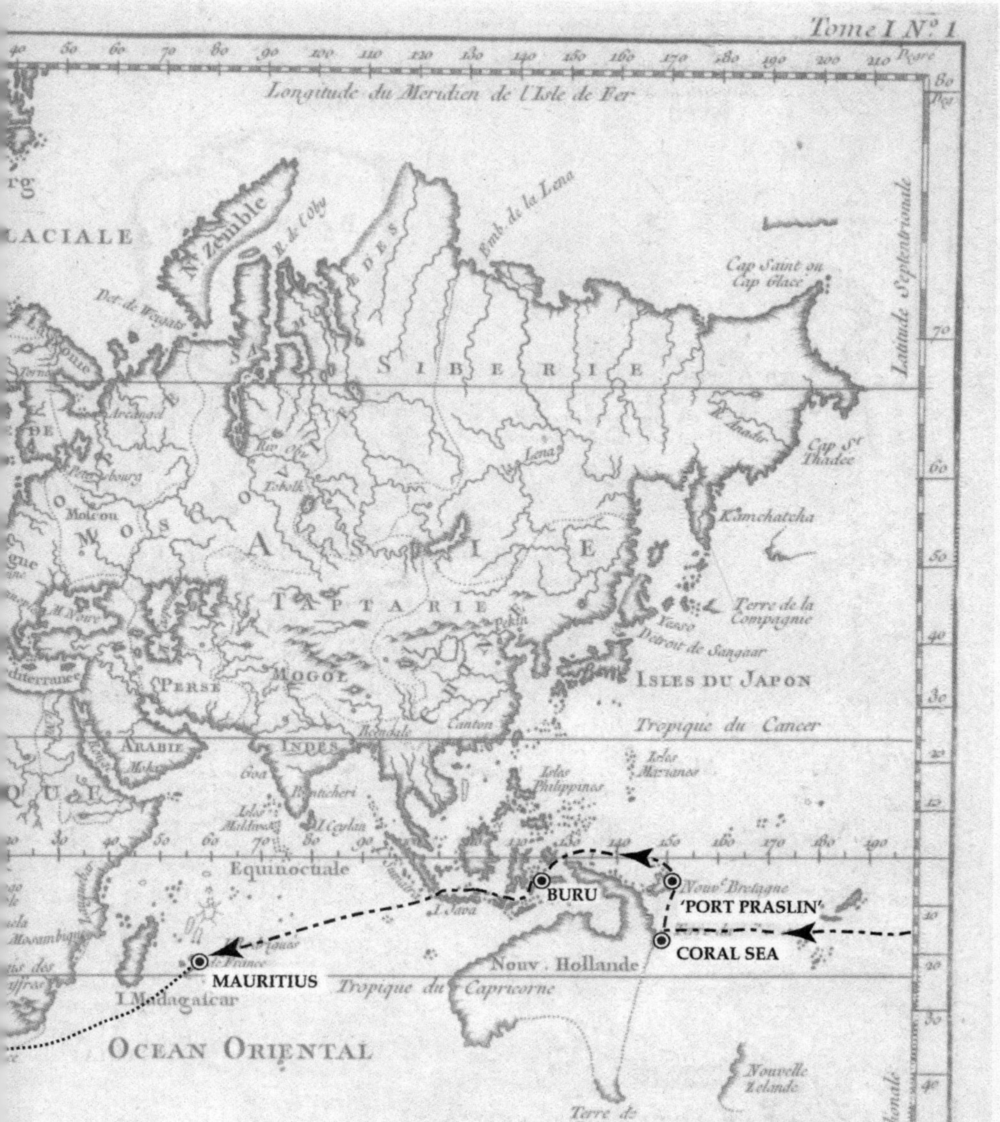
THE VOYAGE OF JEANNE BARRET
------Étoile 1766–1768
..............Sympathie 1776
BURU
'PORT PRASLIN'
CORAL SEA
MAURITIUS
Tome I N°. 1
Longitude du Meridien de l'Isle de Fer
Longitude Orientale du Meridien de Paris
Latitude Septentrionale
Latitude Meridionale
N. Zemble
Emb. de la Lena
Cap Saint ou Cap Glacé
S I B E R I E
Cap St. Thadée
Kamchatcha
A S I E
T A R T A R I E
Terre de la Compagnie
Yesso
Detroit de Sangaar
Isles du Japon
Tropique du Cancer
Perse
Mogol
Arabie
Indes
Canton
Isles Mariannes
Isles Philippines
I. Ceylan
Isles Maldives
Equinoctiale
I. Java
Nouv. Bretagne
Nouv. Hollande
I. Madagascar
Tropique du Capricorne
Ocean Oriental
Nouvelle Zelande
Terre de Diemen
Antipodes de Paris

If we fail to construct our own realities, other people will do it for us
– Epeli Hau'ofa, 'Pasts to Remember', 2008

It is perhaps a more fortunate destiny to have a taste for collecting shells than to be born a millionaire
– Robert Louis Stevenson, 'Lay Morals', 1896

It would be enough; as an alibi for a life, it would do; she would not need to apologize for how she had spent her time on this earth
– Amitav Ghosh, *The Hungry Tide*, 2004

B: She was always wise and brave.
A: These frail machines sometimes contain strong souls.
– Diderot, *Supplement to Bougainville's Voyage*, 1772/96

Something white catches in the surf as the waves roll and retreat across the beach. Like a scrunched-up ball of paper batted between a kitten's paws, it tumbles and spins in the water: caught, dropped, then suddenly tossed back into motion.

A tern skims low on angled wings for a closer look, one dark eye tracking the object along the drifting shoreline. Nothing else moves on the beach. No tracks mark the fine white sands. Nothing stirs in the dunes bristling with tufts of seagrass. The red and gold cliffs warm as the morning sun rises over the hills. The only sound is the irregular wallop of the waves and the hissing sigh of the retreating surf.

The mysterious object is a paper nautilus: not a true shell at all but the egg case of the pelagic octopus *Argonauta nodosa*. A shell more fragile than the finest Venetian milk glass. Nodules fan out in spiralled rows from the apex, tipped in brown along the double-ridged keel. The slightest pressure between two fingers would crush it like an eggshell. How could anything so brittle survive the ocean swells, much less the impact on the shore?

Despite its fragility as it rolls in the surf, the paper nautilus is still intact. Its voyager has long departed, leaving only the most insubstantial evidence of her ephemeral life on wild seas. A frail machine, indeed, to contain such a strong soul.

1

THE CALL OF THE SEA

Port Lincoln, 1974

I ALWAYS DREAMT OF a life at sea. As a very small child – no more than four – I stood in the shallow waters of Porter Bay on the mid-coast of South Australia watching the motley fleet of neglected yachts and gleaming fishing vessels bob on their moorings. I remember puzzling over where you slept on them. I must have thought that their hulls were solid, like the toy boats my dad made from timber offcuts that I floated in puddles. I imagined sliding into the low cabins on the deck, like the narrow slotted bunks in the back of my grandparents' campervan.

My ignorance about boats could not have lasted long. My dad was a boatbuilder and I spent my early years in and around the slipways and boatyards of Port Lincoln, on fishing boats, dinghies and sailboats – our own and other people's. As I played on the beach at Porter Bay, I would design imaginary boats in the sand: a half-buried tree-branch hoist upright for a mast, a blanket of black seagrass for a bunk, crooked lines of weathered stones defining cabins, decorated with shells, frosted glass and leathern sharks' eggs. It was a boat only in a child's imagination.

I took the best of the shells with me, tucked safely in my pocket – the delicate painted ladies and bubble shells, whelks and periwinkles, cones and tiny cowries. The rest would be left to wash away on the incoming tide. That was the best part of the beach – the fact that it washed itself clean, as if no-one else had ever been there.

My dream of a life at sea was not an idle fantasy, nor even a particularly imaginative one. My ambition reflected that of my parents, who had spent several years building a gaff-rigged yawl in the back blocks of Port Lincoln before launching our small family out into the Southern Ocean to cruise the coast of south-east Australia up to the tropics of north Queensland. It all seemed perfectly normal to me. At the time, in the mid-1970s, they told a local journalist that they wanted to 'escape from the constraints of the clock'.

And so, when I imagined my future, I dreamt of continuing that tradition on my own. I imagined myself aboard a flat-bottomed Dutch barge, under the sleek gaff rigs of Thames Quay punts, or in gleaming white couta boats, all of which would be artfully fitted out with cabinetry to accommodate my collection of shells and books.

I decided that I would be a marine biologist. It was the only job I could think of that would allow me to live on a boat. I wasn't exactly clear on how you became one, what marine biologists did, or who employed them. I'd certainly never met one. But the local high school biology teacher gave me a booklet on how to collect and prepare specimens, how to label them and record their provenance. I dutifully identified the species of each new shell I found, numbered, catalogued and stored them in boxes stuffed with cotton wool.

My first 'big' shell was a southern spindle. This classic gastropod is often described as a whelk but is distinguished by the satisfying symmetry provided by its elongated siphon canal. This one was only about 10 centimetres long – nothing spectacular in the scheme of things – but it was rare to find any undamaged shells along the rugged south coast. I still remember the thrill of seeing it lying ahead, elegant and unbroken on a long white stretch of sand. I took it home, cleaned it and compared it with the pictures in the growing collection of shell-identification books lining my small bookcase.

Reading fed my obsession with all things maritime. As an only child with a companionable Siamese cat for a playmate, I had plenty of time for both. The colourful pages of *Scuppers the Sailor Dog* still elicit a Proustian glow of recognition from the inner child yet to embark on her voyage. I longed to live the idyllic shipwrecked life of R. M. Ballantyne's *The Coral Island*, while Willard Price's *South Sea Adventure* shaped my expectations of giant manta rays and man-eating octopuses. And my teenage years were succoured by the endless maritime exploits of Hornblower and Bolitho, Aubrey and Maturin.

Even then I noticed that there was not really a place for someone like me in these stories. Girls did not go to sea. And when women appeared they were relegated to the periphery of the action, rarely occupied a central role, and, when they did, quickly died. Perhaps this did not bother me so much then. Children have a flexible sense of self, readily casting themselves into the role of another. That is, after all, the purpose of literature.

But as I read more, I noticed women in places where they were not supposed to be: stowaways, wives, admirals, pirates, lovers, fleet commanders, whalers, sailors, servants and cross-dressers. I read snippets of stories, mentioned in passing: an aside, an anomaly, an unfinished account. And I remember drawing pictures of classic eighteenth-century naval uniforms, with belts, buckles, long boots and swirling coats – and culottes that a female officer might reasonably wear while still conforming to the romantic style of the era.

I must have wanted a place for women in these stories – a place for myself. Stories construct the mythologies of our history, help us understand our present circumstances and illuminate the possibilities of the future. If we cannot see ourselves in stories, how can we imagine ourselves creating new ones? We all stand on the shoulders of giants in our achievements, but it is difficult to find a firm foothold on the shoulders of those who have been erased from history. The great age of sail, the golden era of discovery, is very much seen as a man's world in which women, if present at all, were an inconvenient burden or a source of discontent. But is that really true?

Gudrid Thorbjarnardóttir the Far Traveller sailed to Greenland and Newfoundland 500 years before Columbus, gave birth to the first European born on American soil, and travelled back to Greenland and Iceland with her family and even as far as Rome. But history rarely mentions anyone other than her shipmates, Erik the Red and Leif Erikson. Women sailed into battle at Trafalgar, alongside their husbands or dressed as men. They worked as nurses and cooks but also served in war, were injured and died. They signed up as sailors, climbed the rigging and were accepted by their peers.

Despite women being banned from ships, nearly every French expedition I researched for *Voyages to the South Seas* seemed to have had a woman on board in some capacity or other – a stowaway wife, a woman dressed as a man, a convict escapee or a captain's paramour. All carefully expunged, reduced or ignored in history. Rose de Freycinet, the wife of the captain Louis de Freycinet, was the only one who left her own account, of her voyage around the world in 1817–20, although her letters home were not published until 1927.

By and large, these women did not tell their own stories, and others did not tell their stories either. I have often wanted to know what happened to young Mary Beckwith, sentenced to death, then life-transportation to Australia, for stealing calico, who negotiated her escape from the colony on board Nicolas Baudin's *Géographe* in 1801. Or Marie-Louise Victoire Girardin, who sailed as a steward on the d'Entrecasteaux expedition in 1793, fought a duel to defend her honour and became the first European woman to visit Tasmania. Or the teenager Louison Séguin, who sailed with Kerguelen in 1772 to the subantarctic isles that now bear the captain's name.

These women drift in and out of the official records, histories and biographies. They are seen as anomalies for which there is too little information to be worth discussing in any detail. Their fates and lives are largely undocumented. And just as little is known about the French woman whose example may well have inspired them and whose exploits and achievements were at least reported at the time, if not in any great detail.

I cannot remember the first time I heard about Jeanne Barret. Perhaps it was when I was researching the French voyages to the Pacific. Maybe I had heard mention of her earlier, in my youthful reading. Nowadays I notice her absence more than her presence – whenever she is not mentioned in discussions of botany, of circumnavigations, or of pioneering expeditions. You would think the very first woman known to have sailed around the world would be noted more often, that her role as a botanist and collector of one of the most significant collections made in the eighteenth century would be worth mentioning.

I have always wanted to know who she was and why she made her journey, and what happened to her on her return. But she left no account of her own. No letters or journals or publications have survived written in her hand or in her own voice. And so what wasn't known about her was filled in with stories and myths: colourful, implausible, romantic and tragic. Jeanne suffered from the same mythologising as the argonaut octopus, which for centuries was said to hoist its tentacles as sails, row them like oars, or steal its shell like a hermit crab. Too often a good story is allowed to stand in the way of the truth.

If the famous explorer and war hero Louis Antoine de Bougainville had not mentioned Jeanne in his bestselling account of the first successful French circumnavigation in 1766–68, *A Voyage Round the World*, her achievements would probably have been entirely forgotten. According to Bougainville, she disguised herself as a man in order to secure a position as assistant to Philibert Commerson, the naturalist. Her true identity and gender were only revealed in Tahiti and she confessed to Bougainville that she was an orphan from Burgundy who had been made penniless by a lawsuit. She had disguised herself as a man to find work as a valet and had signed on for the voyage out of curiosity. Bougainville allowed her to continue in her role until they reached Mauritius, where she disembarked with Commerson.

Even with this brief mention, her story might have died away. She appears, largely, as a footnote in Commerson's life story. Commerson's friend and biographer, the astronomer Jérôme Lalande, did not mention her by name in his eulogy for Commerson published in 1775, other than praising her indefatigable courage and singular discretion. The botanist Jean-Baptiste Lamarck briefly acknowledged her contribution to Commerson's vast botanical collections, her courage and her love of travel, but assumed that she had died on the journey.

Most of the collections of birds, fish, mammals, insects, plants and shells she made for Commerson were lost, neglected or went unpublished. It was as if when Commerson died, so too did Jeanne.

But Tahiti was rich in French Romantic imagination, and a woman exposed on Tahitian shores could hardly go uncommented. The Tahitians saw her for what she was, when her shipmates could not. Bougainville's own accounts of beautiful and inviting Tahitian women fed an interest already ignited by Commerson's rapturous letter published in 1769 about the island where 'neither shame nor modesty exercise their tyranny at all' and 'the act of creating a fellow human being is a religious one . . . the climax celebrated by universal applause'.

For good measure, the philosopher Denis Diderot further inflamed enthusiasm in his fictional reimagining or 'Supplement' to the voyage of Bougainville, espousing greater social and sexual freedom. Jeanne Barret appears in this story too, not in her own right but as a cipher for the corruption and degeneration of modern society. But even as a metaphor Diderot admired her courage in venturing forth.

'These frail machines sometimes contain strong souls,' he concluded.

In 1774, Commerson's last will and testament was published in Paris, making it clear that his housekeeper, Jeanne Bonnefoi, was in fact Jeanne Barret. Suspicions over their relationship, which had simmered on the voyage, were aroused. Over the course of the nineteenth century, a few journalists republished Bougainville's narrative or recounted Commerson's life, embroidering them with suggestive details. The art critic Louis Petit de Bachaumont could not resist the gossip of Commerson's 'genius for a little romance', reporting that he had left his housekeeper in Paris, but that she 'was very attached to her master, whose bed she knew better than the kitchen', following him aboard and (improbably) keeping her identity secret from him for several months.

For the next century, Jeanne's story was filled with speculation and an accumulation of anecdotes and assumptions whose origins and authenticity are unclear. There are too many gaps and absences. The stories feel clichéd and stereotyped – she was by turns a loyal and devoted beast of burden, a courageous and stalwart servant or, more recently, a victim of sexual violence.

All that is left of Jeanne today are a few fragmentary traces in the archives, a handful of documents, a signature here or there, a reported conversation and descriptions from others, some malicious but mostly admiring. Most of the accounts were written or rewritten long after the event, and the only supposed picture of her, in stripy sailor clothes, was constructed after her death by an artist who had probably never met her.

With so little information about Jeanne, it would be tempting – easier – to write a novel to fill in the gaps with imagination. Or to write the history I would like to find, coloured by my own preferences rather than shaped by evidence. But there has been far too much speculation and imagination about Jeanne already.

Like the Tahitians, I feel the need to strip away the assumptions that clothe the historical figure of Jeanne Barret, to see the flesh and bone beneath, the beating heart and thinking brain of the person who lived and breathed but left no words for us to hear. Who was this woman? How did she fit into that male-dominated world of exploration and sail?

I want to follow the advice of the first biographer to investigate Jeanne as a person rather than as an accompaniment or anomaly. In the 1980s, Henriette Dussourd wrote in a small book on Jeanne's life that she 'preferred to leave blank space for other researchers to eventually complete' rather than 'advance anything that is not supported by archives'.

I don't want to create any more misconceptions. I want to try to understand the real woman, as a product of her own times and environment. I am not sure I can successfully strip away my own expectations and values, but I can at least try to critique them as vigorously as the myths and misconceptions that have seeped into Jeanne's story. How else can we understand how an impoverished domestic servant from Burgundy became a trained botanist and successful explorer?

My journey to find Jeanne has taken me several years, around the world and back in time – to France, Mauritius, and the Pacific, to my childhood, and to the eighteenth century, that period of history that has had such important and resounding implications for Pacific colonial history. I am not sure if my childhood at sea or my scientific training has any bearing on understanding the experiences of an eighteenth-century French peasant woman on a ship full of young men. But as others have said, 'most scholars are landlubbers' and historians are not usually scientists, so perhaps I can find some new perspective by looking at the same material in a different light.

It might seem unlikely, after more than two centuries, that there is any new material to find on this woman. But if there is one thing my training as a biologist has taught me, it is that there is always something new to discover if you choose to examine things closely enough. It took the self-taught Jeanne Villepreux-Power to invent the aquariums needed to study the life cycle of paper nautilus argonauts in the 1830s, finally dispelling the ancient myths of Pliny and Aristotle, and replacing them with observations and evidence.

I wonder if there is still something hidden in the archives, or in the collections of specimens that Jeanne Barret spent so much time finding, carrying, preparing and organising. I wonder what others have missed, what has been assumed and taken for granted, what a closer reading of the archives might reveal.

If we want to find a place for women in history, then we have to start studying them. If we want to understand the lives of ordinary people, then we have to search for their traces wherever we can find them, no matter how unremarkable or inauspicious they might seem. And if we want a history that is not simply written by the victors – a history of the well-educated, well-off European men – then we have to start looking at the history in the spaces in between.

There is only one place to start with Jeanne, and for that I need to go back to the beginning.

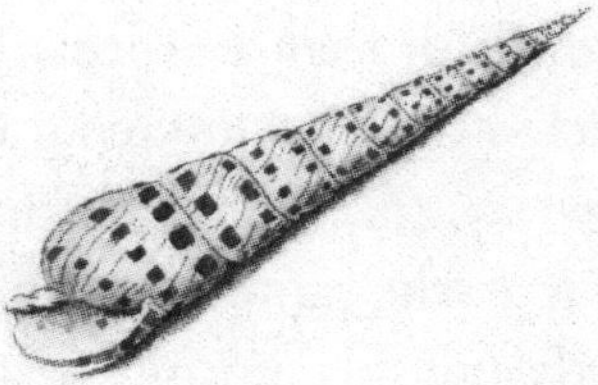

2

AN INAUSPICIOUS BEGINNING

La Comelle, 1740

Jeanne's birthplace is the tiny village of La Comelle, in the southern region of Bourgogne or Burgundy, barely 300 kilometres south-west of Paris. My first challenge is to get there. It's not on a major train or bus route and I'm nervous about driving on the opposite side of the road on my own in a country where I can barely read the road signs. Fortunately, some English friends with a house in the region have kindly offered to drive me around the area where Jeanne was born. I met Miles and Jocelyn at a book launch in Adelaide when they were visiting Miles's daughter, who is also a writer. I think they understand how writers work. They have swiftly organised a perfect itinerary, with side visits to museums, meetings with locals and suitably atmospheric farmhouse accommodation.

My new friends suggest that I catch the train to Saint-Pierre station a day or two after I arrive in Paris, to meet them before we all drive east. Sounds simple enough, although getting a train ticket proves more difficult than I expect. The ticket machine will not sell me a ticket to Saint-Pierre and the man guarding the queue for the ticket office diligently declines assistance to anyone not travelling on the same day. I stoically return in the morning an hour early and this time the man at the ticket office has a friendly smile and swiftly answers my halting French in amused English.

'Oh, that's a Transilien train. You need an Île de France ticket machine, not one for Grandes Lignes. The white ones, not the blue ones.'

Within five minutes I have a ticket. The train trip itself only takes an hour – an easy weekend getaway from Paris, or even for a holiday home from England. The locals joke that English invasion forces nowadays come armed not with archers and infantry, but with real estate pages and a healthy bank balance.

Jocelyn drives us up into the Morvan, the heavily wooded 'Black Mountain', which rises like an island from the fertile farmlands of Burgundy below. We slide off motorways, onto ever-narrowing rural roads enclosed in lush spring forests. Cities give way to towns, to hamlets or even smaller *lieu-dits* – clusters of houses with barely even a church or shop to connect them. At the top of the mountains, grey clouds close in, and even on a fine spring day a brisk damp wind blows across rocky outcrops. The region was once home to the vast Bibracte oppidum, a Gaulish fortress town of 30,000 people, made famous by Caesar and interlinked with a network of fortified trading centres across the Mediterranean world.

We stop to visit the museum dedicated to the site – a triumph of minimalism celebrating an apex of ancient architectural achievement. It's a strange location for a large settlement, on such poor mountain soil. The substantial fortifications suggest unsettled times before Pax Romana. Perhaps it is no wonder the stocky, well-fed ploughmen from the fertile plains below jibed that 'nothing good comes from Morvan – neither good winds nor good people'. The only thing the *Morvandeaux* could do well, they said dismissively, was breed.

Such legacies of an ancient past linger even today, a source of pride and annoyance. The Bibracte oppidum overlooks the valley where Jeanne's village lies, no more than a two-hour walk away. In the eighteenth century, the rural inhabitants of France were described as sedentary, rarely travelling more than a day's walk from their birthplace, fiercely territorial, deeply suspicious of outsiders and knowing nothing of the broader world. And yet the oppidum, with its archaeology of trade goods from across Europe, suggests that such insularity was not always the case. Poverty and hardship made the people in this region strong and hardy, but necessity also made them more mobile. The *Morvandeaux*, unlike other rural inhabitants, were accustomed to leaving their mountains in search of work. And many a *Morvandelle* made a fine wet nurse for Parisian babies. Jeanne, it seems, was born at the nexus of ancient roads that connected *pays*, territories, principalities, kingdoms and empires: a heritage of resilience, mobility with a view that looked beyond the horizon.

The clouds lift as we head east down the other side of the range and the forests open to a vista of the valley below where the village of La Comelle nestles on the mountain slopes.

Jeanne was born in the summer of 1740. It should have been a time of plenty: a time of growth and rich harvests, vines hung low with grapes ripening ruby-red in the sun. The swelling ears of grain should have swayed in gentle winds as olives darkened on gnarled branches and feasting swallows skimmed over ponds. Summer in the heart of rural France, the mountains of Burgundy – surely the best of times, the best of places, for a baby girl to be born?

But the winter had been harsh. Even in Paris the frosts had lasted two and a half months and the Seine froze solid from left bank to right. Temperatures rarely rose above freezing. Winter stretched its icy tendrils into spring and starving swallows dropped dead from the sky. Snow fell in June and rain drummed a relentless tempo. By the middle of summer, the olive trees had died, and the feeble grains and grapes were barely worth the trouble of harvest.

On the eastern flank of Montagne de la Garde, in the shadow of the Morvan, the village of La Comelle suffered from successive poor seasons. It had no reserves to feed its few hundred souls. The fountain of Saint-Claire wept, overflowed, but offered no miracles.

The seigneur of Château de Jeu took his tithes and drank his wine, heedless of his starving serfs. No help came from Paris either. No-one believed that the harvest had failed, that the country was starving. The price of bread doubled; the Parisians rioted and blamed middlemen stockpiling supplies for their own profits. The villagers of La Comelle were abandoned to their fate.

And they were not alone. Half the residents of Poitou in northern Burgundy died that year. Half a world away, 300,000 Irish would die of cold, disease and starvation. It was an inauspicious year to start a life.

But babies take no mind of weather, arriving wanted or not, and this one was no different.

I cannot tell you if Jeanne was gently bathed, coated in pig fat and bound tight from top to toe in strips of cloth, in accordance with Morvan tradition. I cannot tell you if she was laid in a bed of straw and a *bré* worn smooth by the rocking of countless generations. But I do know that the morning after her birth she was hurried to the church to be baptised, while her mother rested. In all likelihood she was dressed in a cloth cut from her mother's clothes and a white baptismal cap, and carried by her godmother, her godfather following close behind. For the priest of La Comelle, Father Pierre, duly noted her advent in an uncertain, poorly tutored hand in the parish registers:

> On the twenty-seventh of July in the year seventeen hundred and forty was born, and on the twenty-eighth was baptised, Jeanne, the legitimate issue of the marriage of Jean Baret, a labourer from Lome and of Jeanne Pochard. Her godfather was Jean Coreau, a labourer from Poil, and godmother Lazare Tibaudin, who are not signing.

Baby Jeanne's chances of survival were slim, but at least she was assured of her entry to heaven. For the next six months, she would lie in her bed of straw unable to move, stretch or explore, safe beneath the blessed cord strapped across her cradle to keep her from straying and safe from all the evils of this world and the next.

In all the books and articles written about her, Jeanne is typically portrayed as having been orphaned young, as having no family to support her, or sometimes even as having been in institutional care. She told Bougainville that she was an orphan and no-one seems to have questioned that.

The biographies about her do not report when her mother or father died, whether she had brothers, sisters, aunts or uncles. I would have assumed that if others had found the record of her baptism in the La Comelle parish register, then they would also have looked for other family members there. It is hardly worth looking again – but I look anyway. The scanned pages of the parish registers are online and it does not take too long to check the records for a small village like La Comelle.

But Father Pierre's scrawling handwriting is not easy to read. Sometimes he annotates the entries with *baptême*, *mort* and *mariage*. Sometimes he does not. Over time, I become accustomed to the structure of entries. It becomes easier to spot the recurring names. I wait for each page to load and magnify, then scan down with my finger, searching for Jeanne Barret and her parents. It takes days to work through the pages. But families matter and I want to know where Jeanne came from.

My mother's family is French. When the Jaunay family sold their champagne mark to their brother-in-law Jacques Krug, my mother's great-grandfather Frank Jaunay decided to take his winemaking expertise and young family to Australia. It was a long time ago. My grandfather remembered ancient aunts with French accents who pronounced their name with a soft 'Show-nay', but everyone today pronounces 'Jaw-neigh' with a long, front-loaded Australian drawl that sounds nothing at all like French.

It cannot have been easy to be a French immigrant in an English colony. Frank soon parted company with the Victorian winemakers at Great Western who supported his immigration. They had no interest in champagne apparently. He travelled west to work as cellar manager for Château Tanunda in the Barossa, before running the Scenic Hotel in the Adelaide Hills not far from where I now live. There are not many French migrants in Australia. By and large, France is a country of immigration, not one of mass emigration. French migrants have a low profile in an Anglophone world.

Yet France played a long and significant role in Australia's European history. French footsteps are readily found in the place names around Australia's coastline, particularly in the south and west. South Australian shores are scattered with French names – of explorers like d'Entrecasteaux, cartographers like d'Anville, naval ministers like Decrès and botanists like Jussieu and Tournefort. Some are even carved into the rocks.

My first encounter with this French history coincided with my earliest sea journeys across the gulfs of South Australia. My memories of our first sea voyages are vague and intermingled: of a thudding engine labouring through heavy seas; of crawling on all fours so as not to be flung off my feet; of worrying that we would not meet my grandparents at our agreed rendezvous; of watching the boat lurch uncomfortably at anchor from the shore. It was an early lesson that the weather waits for no-one's timetable and that not turning up on time at the right place is often the best outcome.

On one trip to Kangaroo Island we visited Frenchman's Rock – a legacy of the voyage commanded by Nicolas Baudin, who circumnavigated and charted the island in 1803. Baudin stayed on Kangaroo Island for three weeks, collecting plants and soon-to-be extinct animals. The account of his expedition included the first complete published map of Australia, the coastline replete with even more French names now replaced with English ones.

I love the idea that my home town has an alternative shadow history, that it might perhaps have been French, and that I might have grown up in Port Champagny instead of Port Lincoln, overlooking the waters of Golfe Bonaparte instead of Spencer Gulf. It feels like some tangible connection with an inheritance stretched thinly across generations and the world's oceans to a distant inland winemaking region in western France, centuries in the past.

A third of all French babies born in the summer of 1740 would not survive their first year. Perhaps if Jeanne had been born to a well-off family, her prospects would have been better. But she was born into the very bottom of French society, the most marginal of all the peasants, in one of the poorest and most oppressed regions of France.

Her father was a *manoeuvre*, a manual labourer, quite literally one who uses only his hands (*man*) for work (*oeuvre*). *Manoeuvres* owned nothing but the strength of their limbs, and they survived only by the health of their bodies. They leased the poorest of hovels, tumbling stone topped with rat-infested thatch, fortunate if they might have a fire at one end and a pig at the other. They slept on such straw as was not eaten by animals. They carved themselves crude wooden clogs and covered themselves as best they could in tattered rags that barely protected them from the weather.

For the main they lived on unhusked barley and oats made into a crude bread, baked once or twice a year in the communal village oven, hardening into bricks that could only be softened in watery soups. Meat, salt and wine were rare luxuries. They harvested wild foods from the countryside to supplement their poor meals. The ancient Morvan forests that spread up the hill behind the village concealed gnarled trees hung with small tart pears and apples, olives, blackthorns and hazelnuts. The dark soils yielded turnips and celery, dill and coriander, scrambled through with wild strawberries, raspberries, blackberries, dewberries and elderberries.

If the peasants were fortunate, they might be permitted to till a small plot of land, grow a few vegetables, raise chickens or perhaps even a goat or pig. Both men and women worked the fields 'mowing, harvesting, threshing, woodcutting, working the soil and the vineyards, clearing land, ditching, labouring'. If the harvest was good, if the seasons were fair, they might earn good pay harvesting grains, picking grapes or making hay. In addition, if the women could spin or sew or take in other work, they might survive from one year to the next. But few lived long. Half of the people in France died before their 25th birthday.

A recent book on Jeanne suggested that her mother was a literate Huguenot from Brittany who taught Jeanne to read and write. But this is unlikely. There is not much reason to think Jeanne Pochard came from Brittany. 'Pochard' is certainly more common in Brittany than Burgundy, but the name can also be spelt 'Pauchard', and this version is more common in Saône-et-Loire, the region in which Morvan lies, than anywhere else.

There is no evidence that Jeanne's family could write and the parish records strongly suggest that they couldn't. Parish records were always signed by witnesses if they could write, no matter how poorly. If they could not write, the entry usually refers to witnesses who 'did not sign'. Few entries in the La Comelle parish registers were signed by anyone other than the priest. The few who did showed varying levels of literacy – some wrote in neat cursive, others printed, while a few laboured over thick large letters like a child's first efforts at writing. Since none of the entries for Jeanne's family were signed, it seems certain that her family was illiterate.

And in any case, Jeanne's mother did not live long enough to teach her daughter to read. Just a few pages on, fifteen months from Jeanne's birth, I find another entry in the parish register.

> On the fourth of November 1741 did die and on the fifth was buried in the cemetery, Jeanne Pauchard, wife of Jean Baret, manoeuvre from La Pome, aged forty-five years.

I did not expect Jeanne's mother to be so old. But I am oddly relieved to find that Jeanne had her mother at least for that first vital year of nurture that is seared into every mother's memory and missing from every child's – the emotional foundation on which to build a resilient future.

The Church of Assumption in which Jeanne was baptised, and the old cemetery in which her mother was buried, no longer stands at the crossroads of La Comelle. It was demolished in the early years of the twentieth century, deemed too low, damp and small for the growing population. It may well have stood for over 800 years but, having lived with the discomforts of ancient monuments for so long, it is unlikely that many in La Comelle mourned its passing. The new church was built over the top of the old graveyard, eternally consecrating the anonymous generations beneath.

It's unlikely that Jeanne's mother's grave would have been marked by a tombstone anyway. The poor could not afford to mark their graves. The only people who cared already knew where their loved ones were buried. Their testaments were transient and left no trace – wood and flowers, prayers and vigils.

The moment of transition from old church to the new was captured in a postcard from 1901, showing the old church with its simple square Romanesque spire in half-demolished rubble before the rising new church with stained-glass and classical bell-tower elegance. The black-and-white photo gives a wartime bleakness to the scene. My modern eye is incapable of seeing the triumph of progress, imagining only a destruction yet to come: battlefields of future wars across Europe, sprawling urbanisation and environmental degradation.

~

The view from La Comelle today still looks like a nineteenth-century painting of Morvan peasant life by Camille Corot. Violet-tinted skies frame cream-stone buildings under steep brown roofs. In the pastures of gold and green, it's not hard to imagine bonneted peasants working the fields, chopping the wood or walking the pig.

Jocelyn parks the car in the church carpark and the three of us step out into the warm sun to admire the view down the valley where rambling creeks trace uneven paths across the neat green fields.

To my surprise, there is little sign of modern life here. There are no motorways, few powerlines, no industrial complexes, no cities or suburbs within eye or earshot. The houses are either old or traditional. Herds of beefy brown Limousin and white Charolais cattle dominate the landscape, their fields guarded by long-standing hedgerows that provide food and shelter for a rich array of wildlife.

This *bocage* – a mix of small fields, hedgerows and woods – clearly supports a healthier ecology than other European farmland that has been converted to large-scale cultivation. Wildflowers bloom thick along the roadsides as forests lush with new spring growth spread down the slopes. Even through the car window I can hear vibrant spring birdsong. Black hawks hang ever vigilant overhead. A large goshawk swivels on its fence post as we pass, keeping us fixed in a sharp yellow eye.

I often struggle to appreciate European landscapes – so many of the plants and animals are weeds and pests in Australia that it's hard to view them with an unjaundiced eye. But this landscape feels intact – not from an absence of humans, as Australians and Americans are wont to conceptualise wilderness, but in some kind of balance. It is a landscape that has evolved and changed radically with humans but, at least on the surface, does not appear over-exploited or disturbed.

The silence stretches broad and deep across the valley. Small voices drift from a distant field, discussing this season's vegetables. On a whim, we investigate a small well-trodden path leading up between two houses to the woods beyond. It leads to a small shrine surrounded by a stone wall. The overflowing dampness at its hollow base suggests that this must be the fountain of Saint-Claire I have read about in the historical records but never imagined finding. Plastic votive offerings verify that the shrine is still used today, just as it must have been in Jeanne's time. It feels like a tangible connection across the centuries.

The water from the fountain spills down a narrow stone drain through a garden on the roadside. Herbs, flowering both colourful and aromatic, tumble down the hill around it. Miles chats to a lady working in the garden, who offers us sprigs of fragrant thyme. I am grateful for his easy sociability, which seems to attract additional information about Jeanne Barret with enviable ease. Jocelyn thoughtfully translates for me when she sees I have lost track of the fast-paced French conversations. We learn that there are still Barrets living in nearby villages. A road across the way leads up to the village hall, both of which have been named in honour of their famous resident.

As we breathe in the herbal aromas of thyme crushed between our fingers, it is impossible to reconcile this charming village with the grinding poverty described in the history books – the elderly begging for alms or the children forced to eat grass, the deaths and disease, the starvation and hardship. I cannot imagine farm workers so poor that they were forced into a kind of hibernation through winter to save on fuel and food. Come autumn and the labourers of the fields would 'spend their days in bed, packing their bodies tightly together in order to stay warm and eat less food. They weaken themselves deliberately.'

What I experience today as calm, peaceful and idyllic might, in Jeanne's day, have been simply the limits of a life so close to death as to be almost unimaginable.

I think we always love the landscape of our childhood. In Paris, a friend takes me to visit the apartment he grew up in in the 1950s. He is meeting a builder to arrange renovations to the tiny two-room home, which is still in original condition with a toilet in the kitchen cupboard and a shower in the wardrobe. The rooms are so narrow that the large bright windows on the fourth floor give me an uneasy sense of vertigo as we look out over the tiny park below.

'What a perfect place for a child,' my friend enthuses, sweeping his hand over the view, 'with all this to play in.'

I grew up with the wheatfields of South Australia stretching to the north and the vast expanses of the Southern Ocean to the west and south. My parents raised the grey steel frames of our boat in these hinterland paddocks. We lived in a caravan on a friend's bush block, with a portable toilet and a bucket full of holes for a shower. In summer the mice invaded the caravan, soon followed by a 1.5-metre long brown snake that slipped down the back of the cutlery drawer as we prepared to set the table. Tiny ants launched endless campaigns against our supplies, trekking long marauding lines across the ceiling. And in winter, the ice in the rainwater tank froze the tap solid. It cannot have been a comfortable place to live, but for the most part my memories are dominated by the spring wildflowers, the abundance of lizards, birds, possums and kangaroos, the rope playground under the spreading shade of two vast gums, and the magic fairylands of moss and mushrooms under the paperbarks. It was a wild wonderland filled with dangers and adventure, and I had it all to myself.

Not everyone loves the landscape of Eyre Peninsula – wide bare grasslands, salt flats and coastal heather, the stunted mallee vegetation blackened with traces of last season's fires, or rugged coast pounded constantly by vast thundering swells. It is a landscape that can also feel empty, exposed and isolated. The summer sun shrivels everything with oven-like ferocity, while winter gales roll in from Antarctica with monotonous violence.

'Only a South Australian could love this,' my father enthused on a trip home after years living in the wet lushness of northern New Zealand.

I wonder if Jeanne thought the same thing of her childhood home – if she remembered it fondly and with happiness, or if she only remembered the hunger and hardship.

The winters continued cold and punishing through Jeanne's youth, but ever so gradually the weather began to turn. The frosts were less severe, the snows less frequent. Winter returned to its regular schedule.

Someone looked after Jeanne through those early years. If not her father then perhaps her godmother or a stepmother. Jeanne's father soon had a new wife, Antoinette Mangematin, but she, and their son Simon, both died in 1745.

In Jeanne's sixth year, the grapes were still late and the grain harvest low, but the vegetables flourished and filled many hungry bellies with an unexpected abundance. At six, peasant children were expected to work: looking after animals, helping in the fields or at home. They cast off the simple smocks of childhood and put on cut-down versions of adult clothing – jackets and pants for boys, a short skirt and bodice with a white cap and apron for girls. They could earn money, food or clothes by working on local farms, or winding bobbins for weaving, or as domestic servants, advertising their availability at the annual hiring fair.

By the age of seven, Jeanne had lost another stepmother. Jeanne Teuvenot, aged 40, died in November of 1747.

It is some time before I find anything more of interest in the parish records. I've nearly given up, but feel that I should, at least, finish the volume. Near the end, I find another entry:

> On the 16th of December 1755 did die and on the 17th was buried in the cemetery, Jean Baret, manoeuvre aged about sixty years . . . in the presence of his son Pierre Baret, his son-in-law Antoine Gigon and others who do not sign

So it is true, what Jeanne told Bougainville on the ship in the Pacific, that she was an orphan. But she was not alone. She did have family. She had a brother Pierre and a married sister.

I keep searching back through the registers before Jeanne's birth, hoping to find the marriage of her parents or the birth of her siblings, but I find nothing. Perhaps they moved from another village, perhaps I missed the entries, or perhaps the pages were damaged by damp and the passing years. There are entries that might be related, but a name is wrong, even allowing for phonetic spelling, or the date doesn't add up. I can't be certain they refer to the same family.

Finally, in 1734, I find a record that seems convincing.

> The 12th Feb 1734 is baptised Pierre Barret, son of Jean Barret and Jeanne Pauchare the mother, the godfather is Pierre Petit Jean, sharecropper of Poil, and godmother Anne Dularet

This must be Jeanne's brother, Pierre, who would have been 21 when their father died. And then there was the sister whose husband was present at Jean Barret's death. Jeanne was not alone in the world – I am sure these siblings must have been important to her.

By 1764, Jeanne had left her home village. She was not like the childhood village friends of the famed French author Colette who also grew up in Burgundy. Colette described the children of the village as being 'chained to their parent's shops', victims of the 'resigned wisdom, the peasant terror of adventure and distant travel'. Colette herself dreamt of being a sailor.

In a world where few travelled far beyond the confines of their own village, and fewer still trusted those who did, Jeanne picked up whatever meagre possessions she had and struck out to follow the Arroux River south towards the long flat plains below.

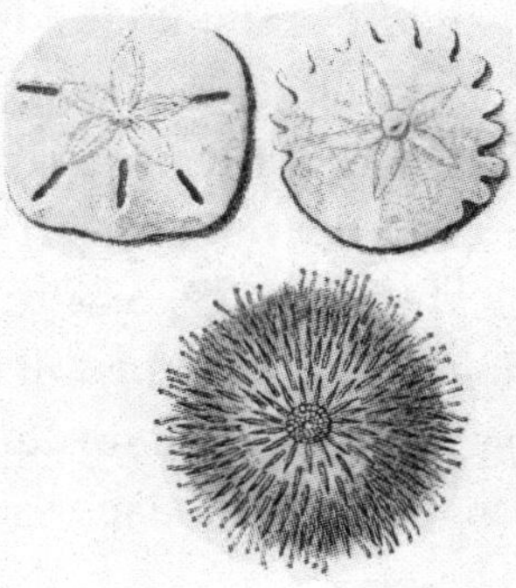

3

MAKING HER MARK

Toulon-sur-Arroux, 1764

IF YOU SIT BY a stream in the woods near La Comelle and the sun hits the water at just the right angle, you might notice an all but invisible shoal of tiny fish, quivering like magnetic needles in the current. They hang as if pinned by their dark eyes, each pointing in the same direction, identical in size and shape, swimming against the current to stay in exactly the same place.

They are glass eels or *civelles*. These mountain headwaters are their nursery. Here they live, feed and grow for up to 25 years, changing to elvers, yellow and silver, before letting go of their childhood haunts and beginning their long migration to the sea.

No-one knows quite what makes them leave: some instinctive urge, some restless momentum propels them downstream, heedless of the risks and perils ahead. After years of holding their place in their one small pond, surviving the challenges of daily life, they suddenly answer the call of a far distant ocean.

And so they give in to the force of the water that they have grown up resisting. They turn tail and tumble down rivulets swollen with winter rains, evading the still ponds of farmlands, before joining the waters of La Braconne until finally they drop unceremoniously into the broken shallows of the Arroux River.

At some point, Jeanne too left her home town and headed downstream, following the river, until she reached the town of Toulon-sur-Arroux.

No-one knows when Jeanne arrived here, why she moved, where she lived and worked. From the time of her birth in 1740 until 1764, virtually nothing is known about her life. All I know is that on 22 August 1764, when she was 24 years old, she was working as a domestic servant in Toulon-sur-Arroux and she was pregnant.

Voyages, like books, can be a long time in the making. I am sure my parents told me it would take a year to build our boat. On the last day of school, I rushed home excitedly, thinking that everything would be miraculously complete and that the skeleton of bare steel frames would somehow have transformed into a completed vessel ready to launch. It had not.

We moved into the half-built steel shell, clambering up a homemade ladder. My parents slept in the two bunks in the focsle while I camped on a makeshift bunk over the chain locker, with a cardboard box for a cupboard between the hawse pipes in the bow. The route across the still unsealed main section to the galley in the stern where we cooked and ate was a labyrinth of steel beams and gaping holes, best navigated with speed and enthusiasm. It was like living in a cubby house with a built-in adventure playground. I think I only fell once: I barely remember the graze on my thigh.

Eventually the hull and deck were sealed and the name *Caliph* welded in steel letters across the stern, in the days when Islam was synonymous with nothing more ominous than Eastern exoticism. Everything was sandblasted and zinc-treated, painted with metal primer then sealed with dashing black-and-white topsides. *Caliph* was loaded on a trailer and made a slow and careful trip under low-slung powerlines into town, down to the coast and onto the slipway, accompanied by an excited local reporter.

For the next year or so, we lived up at the top end of the slipway, overlooking the constant movement of fishing boats and yachts up and down for repairs, pungent with oil and paint, shaved wood and hot metals ground by the shrieking whine of power tools. I spent the summer developing an intimate acquaintance with the inhabitants of my favourite rock pool at a tiny beach no-one else used. Over the summer I filled the pool's sandy bottom with 363 tiny dogwhelk shells I collected from the shore and then left them for the winter storms to redistribute.

Jeanne herself may not tell us how she came to live in Toulon-sur-Arroux, but her family is more revealing. The records have already revealed that she had at least two siblings, close in age – Pierre six years older and a sister just three years older. Confusingly, this sister was also called Jeanne. I worry that the record is wrong, that these two people are somehow one and the same. Why would a family give both surviving daughters the same name? I email a French genealogist who has documented this branch of the Barret family to ask if such multiple names are unusual.

'It was common for brothers and sisters to have the same first name,' Alain reassures me promptly. 'For example, with one family all or some of the boys were called Jean and the girls Marie.'

My imagination fails me when I try to imagine a conversation in the Barret household, between the parents Jeanne and Jean, discussing the birth of a new baby sister Jeanne to four-year-old Jeanne. It does not make for an intuitive narrative structure, then or now.

Jeanne was fourteen years old when her older sister married Antoine Gigon in the cold damp church at Dettey, a village half a day's walk away from La Comelle. Such marriages were the product of long and complex courtships, unions between villages to be disputed and agreed upon, criticised and celebrated. The invitations were offered, the bans were read, bad omens avoided. The procession would have made its joyful way to the church with music and merrymaking, women in the lead and a boy with the sacrificial black hen bringing up the rear. And then to feast and to dance, with bread dipped in sweet wine and as much food as the barnyard could provide.

The newlyweds did not stay long in Dettey. Eleven months after the wedding, Jeanne's nephew Romain was born in Thil-sur-Arroux, just across the valley from Dettey. As the old joke went, there were no knots in the *aiguillettes* that tied up Antoine's pants.

Perhaps Jeanne went with her sister to Thil-sur-Arroux to help with her newborn nephew, with whom she would retain a close connection. Or perhaps, as the youngest unmarried sister, she stayed in La Comelle with her father until he died at the age of 60 – a good age for a labourer.

By April 1756 Jeanne's brother, Pierre, married Anne Lanoiselée in the church at La Comelle, bringing a new wife into their home. Another reason for a younger sibling to leave and go help her sister with her growing family.

I can imagine them walking across the fields between these villages, when I look at the nineteenth-century Romantic peasant paintings in the galleries of the Musée d'Orsay. The women wore open wooden shoes and tied a cloth apron over their canvas skirts and short-sleeved bodices. A small scarf was all that usually protected their shoulders. But when travelling, the women covered themselves with a hooded cloak that hardly differed from the ancient cloaks of the Gauls who walked the same roads from Bibracte to the Loire.

Jeanne's brother-in-law, Antoine, died in 1757, when she was sixteen, and it was another two years before Jeanne's sister married again, this time to Jean Lanoiselée, who may well have been the brother of Anne. Over the next few years, Jeanne's sister added three more children to her family: two daughters, Lazarette and Françoise, and a son, Léonard. For a time they lived in Thil-sur-Arroux, but Jean Lanoiselée's employment must have taken him further afield for, on the death of their eldest daughter Lazarette at age seven, they were resident on farmland located inside the fortified walls of the nearby village of Montmort.

At some point the family moved south to Rosières, a quiet farming area on the outskirts of Toulon-sur-Arroux, on a bend where the river runs slow and wide. Jeanne was close to her sister's family, particularly her nephew Romain and younger niece Françoise. I suspect she was living with them in Rosières. This, I think, explains how Jeanne came to be living and working in Toulon-sur-Arroux.

When I look at the map of the movements of Jeanne's family over these few years, I cannot help but see the pattern with a biologist's eye. These movements suggest a home range to me – a small, interconnected family moving back and forth along the narrow 20-kilometre corridor of the Arroux River valley. There are Lanoiselées as well as Barrets buried in the current La Comelle cemetery. The women are dispersing out, into the families and villages of their husbands. It can't be a coincidence Jeanne ended up in the same place as her older sister.

It is no more than a fifteen-minute walk across the fields of Rosières to the busy town of Toulon-sur-Arroux and into the old church at the crossroads where the local priest, Father Beau, delivered Mass in the 1760s. In 1762, Father Beau had just buried his own beloved sister, Marie-Antoinette-Vivante, who died after childbirth. Her bereaved husband, the local doctor, Philibert Commerson, was inconsolable and ill equipped to care for his infant son. He needed a *gouvernante* – a housekeeper suited for a widower or bachelor.

This is the Philibert Commerson with whom Jeanne would travel the world. This too cannot be a coincidence.

I have been avoiding introducing Commerson. He was not an ordinary man by anyone's standards. Commerson has comprehensively left his mark on the pages of history and I know that his voice will overshadow that of Jeanne's. I do not want this to be his story, no matter how interesting he was, but he does require an introduction. Perhaps he was not much to look at – slightly taller than average, just over 170 centimetres, with large black eyes and an aquiline nose. But even in the archives he is larger than life and has much to say, as others have much to say about him. His friends described him as erudite, knowledgeable, ardent and passionate, enthusiastically engaging with anyone he found interesting or learned, whether they were a savant or a servant, but contemptuous and dismissive of anyone he found dull or untrustworthy. Jeanne, we can safely assume, was neither of these things.

Commerson became the local doctor in Toulon-sur-Arroux after marrying the well-off widow Antoinette Beau, seven years his senior. He was already 30 years old. Like many early naturalists trained in medicine, his pursuit of science must have seemed like a waste of time to a conservative family when he could have earnt a steady income as a medical practitioner. This wealthy widow, who shared his interest in botany and provided a house and a position as a doctor in her home town, must have seemed like a perfect match.

Commerson was quite clear about what he expected from his new partner.

'She is something of a philosopher,' he wrote to a friend, 'Of a mature age who, through a fortunate conjunction of circumstances, has pleasant features, a good mind and a knowledge of literature, and finally, among her minor attractions, enjoys a fortune of 40,000 francs, most of which is already at her disposal. I do not think that, by marrying her, I am changing my own condition, because I am sure I will be able to make her share my tastes. I have already given her a taste for natural history, and our walks have become real botanising expeditions.'

But the couple had only two years together before Antoinette died three days after giving birth to her first child, at the age of 42, on 19 April 1762. Commerson was devastated by her death, and his brother-in-law, Father Beau, seems to have stepped in to arrange for Jeanne to assist Commerson with his house, his work and his surviving infant son Archambaud.

A recent biography of Jeanne suggested she must have been a herbwoman, that it was her skill with plants that attracted Commerson to her. It's an appealing theory and it gives Jeanne an attractive sense of agency in their relationship – that she taught the botanist botany rather than vice versa. I would like to believe this is true, but I can't see any evidence for it.

The biography suggested that a table of medicinal plants among Commerson's archives, documenting herbal remedies for common ailments, is Jeanne's work rather than Commerson's. Jeanne almost certainly knew a lot about plants, like most peasants of the time. But it is also exactly the sort of thing a country medical practitioner had to be skilled in. In fact, Commerson was famous for his medicinal herbal 'Swiss tea'.

I retrieve the table of medicinal herbs from Commerson's archives when I visit the Muséum National d'Histoire Naturelle in Paris, but I can't be certain that the handwriting isn't Commerson's. It might have been written by someone else or it might just have been a younger Commerson. It has the same backward hook on the *d* sweeping over the letters in front, the same looping *l*, the same symmetrical *f*, the same loose, casual style.

And I can't compare this document to Jeanne's handwriting. The only thing Jeanne is known to have written is her signature. Seven letters are not enough for a handwriting comparison. There is no proof that Jeanne wrote the table of medicinal plants, nor any proof that she didn't. It remains an attractive, but unsubstantiated, theory – useful for a novel but not in nonfiction.

Jeanne's signature, small and rare though it is, does reveal some important information, however. It suggests that at some stage between her father's death in 1755 and her earliest signature in 1764, she learnt to write her name.

The letters of Jeanne's earliest signature are straight, well spaced and clearly formed, like those of a capable student who is proud of her newly acquired penmanship. There are no unnecessary flourishes or ornaments. This is not a laboured and poorly formed signature, like the peasant names inscribed in the parish registers by hands more accustomed to a hoe than a pen. Nor is it the hasty scratching of someone who writes a great deal, in a hurry, and who takes their literacy for granted. Her writing is good enough that it seems likely she may have learnt to write and read more generally, although I cannot be certain.

The capitals of her name fall below the line of writing – not unexpected for *J*, but unusual for *B*. I once read somewhere that, from a psychoanalytic point of view, capital letters signify the ego, that they represent the perception of self, aspirations and self-esteem. If I were inclined to believe in such things, I might be tempted to think Jeanne's signature reveals that she subjugated her own desires to those of others, that she sought to fit in, not stand out from those 'others' represented by her lowercase letters.

Philibert Commerson's signature is strikingly different. He used the old-fashioned long or medial *s* in his name – which looks like a cursive *f* when handwritten. It is a letter that disappeared in French print between 1782 and 1793 but was retained in handwriting into the late 1800s. The habits of handwriting are slower to change than a typeface on a printer's press. But Commerson's signature is conspicuous for more than this feature. He begins with an extravagant flourish – a *C* that completely encases the rest of his name. One might speculate that his ego was all-encompassing – he overpowers and overwhelms all others. It's an entertaining metaphor.

Some psychoanalysts also believe in nominative determinism – that we gravitate towards the meaning of our own names. A professor of French literature observed that the origin of Jeanne's surname is *barre* in French – a bar, a block, a barrier. In writing it refers to crossed-out text. And at sea it describes the horizontal lines of the handrail and the tiller. The verb, *barrer*, is to take the helm and to steer the course. In all senses, it is a strikingly apt name.

More significantly, though, Jeanne's signature tells us how she spelt her name. The local priest in La Comelle, Father Pierre, always wrote Jeanne's family's name as Baret. Bougainville always spelt it Baré. They all sounded identical in the pre-revolutionary, non-standardised world where French was not even the universal language of France and what mattered was not the spelling but the sound. Her mother's name records the same phonetic variation: Pochard, Pauchard, Panchare. Perhaps it is significant that the Barret descendants still living in the area also spell their name with a double *r*.

Jeanne's signature comprises the only words I can be sure that Jeanne wrote herself. In time, a few who knew her would report her words, others would speak on her behalf, but most would stay silent, or ignore or erase her from their stories. In an account where she is almost entirely without words, we have to accept the only words we can be certain she actually wrote and spell her name as she wrote it: Barret.

On 22 August 1764, Jeanne travelled 27 kilometres south from Toulon-sur-Arroux, along the river to Digoin. She went to visit a notary to make a legal declaration. There were plenty of notaries in Toulon-sur-Arroux – a full day's walk seems a long way to go to have a legal document signed.

Jeanne was 24 years old. She could be described as 'small of stature, short and plump, wide-hipped, shoulders in keeping, a prominent chest, a small round head, a freckled complexion, a gentle and clear voice, [and with] a marked dexterity and gentleness of movement'. Some would say she was neither ugly nor plain. Perhaps she was like Jane Eyre, 'poor, obscure, plain and little' but, like Jane Eyre, not to be underestimated.

The notary wrote out Jeanne's statement in a practised hand and Jeanne signed her own name neatly beneath. Her witnesses were neither servants nor farm labourers, but distinguished gentlemen – unexpected companions for a domestic servant.

> Jeanne Barret, adult daughter of the late Jean Barret and Jeanne Pochard, employed as a domestic servant, in the town of Toulon-sur-Arroux declares in writing to His Majesty that she is about five months pregnant.

It was Henriette Dussourd, a historian from Toulon-sur-Arroux, who unlocked the secrets of Jeanne's past, who identified Jeanne's birthplace and found her declaration of pregnancy. Dussourd's biography, published in 1987, was the first substantial work to examine Jeanne's life in her own right and the last book Dussourd wrote. She briefly mentioned Jeanne in 1964, in a chapter on Commerson for a book on Toulon-sur-Arroux. I cannot begin to imagine how she found the record of Jeanne's christening, without knowing which parish archives to search. Nor how she found the declaration of pregnancy among the 3.6 kilometres of notarial records in the Saône-et-Loire departmental archives. However she did it, Dussourd's work provided the first connection between Jeanne and Commerson in Burgundy.

Dussourd made a thorough search for additional proof of a connection between Commerson and Jeanne, but found none. Nor did Commerson reveal his secrets. He was rarely straightforward in his writing, except about biology. He was loquacious and verbose, yet secretive and obscure. He talked in riddles and metaphors. He never mentioned Jeanne by name in his letters, even when she seemed to be the subject of them. I read his letters over and over and just when I think they are making sense, they slip away from me, hinting at events and emotions that I can only guess at.

In a letter to his brother-in-law, Commerson seemed to associate the arrival of Jeanne in his house with a particularly miserable time when he had never felt 'less valued either morally or physically'. I have no real idea of what he meant, but he dated this to 'the deplorable Eclipse of the Star who had led me there'.

Commerson was constantly heading off on expeditions, falling ill and getting injured. He was famed for once having fallen into a ravine and been left, hanging by his hair, until he tore it out to escape. On another expedition he was bitten by a dog; the wound turned septic and continued to ulcerate and reinflame throughout his life. His brother-in-law believed that Commerson's passion and genius for natural history drove him to excess.

'He often spent 15 and 20 days and as many nights in a row without sleeping and without a moment of rest to study, observe and write,' recalled Father Beau. 'He allowed himself just a few moments to hastily take coarse foods and eat only bread, vegetables and cheese to give more time to his passion for the Sciences which did not provide him any resources for the needs of life, and to fully satisfy it, he sold an estate of about fifty thousand livres that was almost all his heritage. For 8 or 9 consecutive years he spent the summers alternately in the Alps and the Pyrenees to look for plants and insects in these mountains that he travelled 3 or 4 months in a row; he lived on bread and milk that he bought from shepherds and slept in their huts on foliage.'

It would not surprise me at all if Commerson had raced off to watch an eclipse of the sun and fallen ill as a result. Perhaps it was the famous eclipse of April 1764, which swept over north-western France, its path documented in exquisite detail by Commerson's friend the mathematician Nicole-Reine Lepaute. I asked a colleague who works in French history for her reading of Commerson's obscure language.

'I don't think it's an actual eclipse,' she explained. 'I think it's a metaphor. It's "the star", a particular star, person or thing that led him here, not the sun, that is eclipsed.'

The star that led him to Toulon-sur-Arroux was Antoinette, his wife. The deplorable eclipse must refer to her death. Commerson was bereaved of the woman he regarded as his soul mate, his fellow naturalist. 'I have lost . . . the most tender and virtuous of spouses, and I only exist today through the memory of having belonged to her.'

He mourned his wife throughout his life. He named a plant with two hearts, two seeds in its fruit, with their surnames *Pulcheria* ('Beau' in Latin) *commersonia* after their union. And while he asked that his body be donated to science, he wanted his heart buried next to his 'always dear' wife in the parish church of Toulon-sur-Arroux.

Perhaps Father Beau did not know how else to support his grieving and profoundly depressed brother-in-law, so he offered him practical support in the form of a hardworking servant, Jeanne. He could hardly have predicted how that would turn out.

I suspect that Jeanne, and the unnamed father of her child, would have much preferred to have kept her pregnancy a secret. It might seem strange, then, that she would make a legal declaration about it. But pregnancy in eighteenth-century France was regulated by the state. In 1556–57 Henri II had decreed that:

> Every woman who is equally surprised and convicted of having guarded or concealing both a pregnancy and a birth, without having declared neither one or nor the other one, and having appeared of one or of the other sufficient testimony, about the life or death of her child after having gone

> out of her abdomen, and after the infant having been, so much deprived of the Holy Sacrament of the Baptism as of the public accustomed grave, such a woman shall be convicted and renowned of having committed homicide unto her infant, and for repair, she shall be punished by the death sentence as her last suffering.

In effect, the penalty for failing to declare a pregnancy was death. This edict was regularly refreshed through the centuries, well into the eighteenth. All illegitimate pregnancies were to be declared through a legally signed *déclaration de grossesse*. Failure to do so could be taken later as evidence of infanticide if the child died or disappeared. In part, this edict appeared to be part of a campaign to enforce paternal financial responsibility for illegitimate children to reduce the burden on the state, but it did not always seem to achieve its goal.

Jeanne made her declaration at the last possible moment – at five months – when her pregnancy would have become difficult to conceal. If Jeanne's tracks are difficult to trace, perhaps it is that they have been deliberately obscured. Sometimes silence can tell you as much as evidence. Just as Commerson never mentions Jeanne in his letters, Jeanne never mentions Commerson in her declaration of pregnancy. She declared her pregnancy because she had to, not necessarily because she wanted to. Two weeks later she departed for Paris with Commerson. They left behind Commerson's young son in the care of his uncle, Father Beau, and once they arrived, Jeanne took on the name of Jeanne de Bonnefoi. There is no written trace of any association between Jeanne and Commerson in Burgundy. There is nothing to even connect her to Commerson's former life, nothing to suggest that he is the father of her child.

But what other reason could there be for Commerson taking a five-month pregnant servant to live with him in Paris?

It smacks of small-town scandal. The evidence is all circumstantial and coincidental – and yet it is compelling all the same.

It seems almost certain that Commerson must be the father of Jeanne's child.

I pick up a small white disk from a market stall, imprinted with a familiar five-petalled rosette. It is a *Scutella* or sand dollar. The fossil is smooth and shiny against the rough matrix that encases it. Sea urchins like this are abundant in the Miocene limestone strata of the Loire Valley, stranded from a receding ocean some 17 million years ago. France, Britain and Ireland may have been joined in a single landmass, but the coast of France at the time was inundated from Bordeaux to Brittany, the sea stretching far inland into the Loire Valley. It was not the first ocean to sweep into Europe. Over eons, incursions have come from all directions, creating 'inland' seas, making islands of continents, dividing and conquering landmasses long before human empires began their momentary occupations. Dig beneath the Roman ruins, the Iron Age relics, the stone arrows and clay pots, and you find layers of creatures that tell the stories of ancient oceans and the ever-changing face of the earth.

It is hard to imagine the rounded sand dollar as a living creature, a sea urchin with its velvet pelt of tiny feet rippling in the tug and pull of intertidal sands. Fossils, like archives, require effort and close scrutiny to reveal their secrets.

'It will bring you strength and prosperity,' promises the stallholder who sells it to me. 'Here, hold it. You can feel it letting go of your trauma.'

I am not sure I have any trauma to release, but the fossil urchin exerts a powerful attraction, as they have for centuries. They have been found adorning a Bronze Age gravesite, and have been used as an antidote to snakebite, to cure seasickness and to protect from lightning. They lined the windowsills and dairy shelves of rural cottages to ensure that a family had ample bread, that the milk did not sour, that the cream was rich and safe from witchcraft. They carried special significance for women. Until the late nineteenth century, French wet nurses carried these sand dollars as a talisman to stimulate milk flow.

I wonder if Jeanne took one with her. I wonder if she put it on a shelf in her room. Perhaps her sister gave her one when Jeanne left for Paris, five months pregnant – the only tangible assistance she could give her younger sister as she ventured forth to have her first child in a strange city with no support.

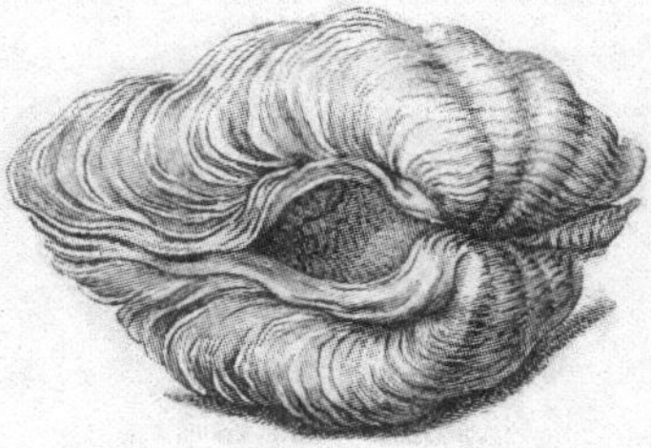

4

A WOMAN OF THE WORLD

Paris, 1764

Living on the far side of the globe, with family and work commitments, I do not get to Paris very often and I do not always get the chance to do the research I would like while I am there. But this time is different. I have a month in France to complete my research in Burgundy, Paris and Brittany – the longest I have ever been away from my family. Just getting there eats up precious time. Even the quickest flights take 24 hours, flying at 900 kilometres per hour. It feels like an eternity and I binge watch a miniseries of Victor Hugo's *Les Misérables* to pass the time.

In 1764, the journey from Toulon-sur-Arroux to Paris took considerably longer and was much less entertaining. Four days with good fortune and fair weather. Not even the pretty little villages and neat vineyards viewed from the well-upholstered comforts of a six-seater diligence offered any relief from the fatigue, boredom and cold. Wooden wheels bounced aching bones over potholed roads, and the only opportunity passengers had to relieve their swollen legs was when they had to march through the snow alongside the laden carriage.

The diligences stopped regularly at inns along the way, allowing travellers and horses to eat, drink and sleep. The food and hospitality varied from exceptional to atrocious. By the time they reached Charenton, where coaches from Burgundy waited for customs approval to enter Paris, the travellers were grateful to step wearily from their carriage.

For a naïve provincial visitor like Jeanne, Paris must have been a revelation: a city of over half a million souls crowded along a wide river. Warehouses packed the banks of the Seine. The towers of Notre Dame rose heavenward in the distance, and everywhere there were buildings and people: a sea of humanity, rising and falling, ebbing and flowing, through cluttered streets.

The city was marked by extraordinary wealth. Prosperity poured from fountains, palaces, gateways and façades. The well-to-do promenaded in their fine gilded carriages and glossy steeds. But those not blinded by gold must have noticed how the river, roads, shops and ships hummed with desperate industry. Vendors threw their hoarse cries for apples, coffee and firewood into the air, hoping to catch some coin. Washed into every corner and every doorway were the discarded, the damaged and the forgotten – the elderly, the sick and the injured. The voices of a hundred provinces all mixed and merged into one Parisian patois, filling the narrow streets of the Île de la Cité, the markets of Les Halles and the laneways of Faubourg Saint-Antoine with their unheeded misery.

Once cleared to enter the city, the travellers crossed the river towards the Left Bank, passing into the quieter, leafy streets of the parish of Saint-Sulpice, near the royal gardens and the Sorbonne. Construction had already started nearby on another grand domed church destined to become a Parisian landmark, this time dedicated to Sainte-Geneviève.

Commerson rented a second-floor apartment here, from a Monsieur Legendre, in Rue des Boulangers – number 15 on the second floor. Old maps and a painting for a later resident reveals that the top floors on both sides of the street had sweeping views across gardens to the north and south. Ground-floor residents enjoyed access to a large rear garden. Jeanne would have collected water from the well in the sandstone paved courtyard, and bought bread from the bakers who gave the street its name, and household supplies from the small shops and street stalls that crowded the narrow lane below.

How did a pregnant young peasant woman from the provinces find life in the middle of such a big city? Did she struggle to understand, to be understood, to lose the *Bourguignon-Morvandiau* dialect she had grown up with? Or had she already heard stories from Morvan wet nurses who suckled Parisian babies at their breast? She could have heard enticing stories from one of the Foundling Hospital agents patrolling the countryside in search of suitable nursing mothers for the wealthy of Paris. Jeanne may well have known a great deal about Paris before she arrived.

Perhaps she was not so much escaping from scandal in Morvan, as arriving in a city of opportunities.

It takes time to reach a point of departure. These things do not happen overnight. Even after *Caliph* was launched and afloat in the bay, it took some years before my family's voyaging began.

Caliph was inspired by Joshua Slocum's *Spray*, an old gaff-rigged Chesapeake Bay oyster boat that Slocum rebuilt from a derelict state and ultimately used to become the first person to sail single-handed around the world in 1895–98. His choice of vessel proved inspired: incredibly easy to handle, stable in rough weather, comfortable and roomy. Most importantly, the *Spray*'s long keel and unique combination of sail plan and lines meant that the boat held its course without the need for constant alteration by the helmsman – perfect for a solo voyage.

Like Jeanne, Slocum is known for being a first in circumnavigation, in his case sailing alone. Like Jeanne though, the full picture is often missing, the rugged male solo sailor eclipses the family man of earlier voyages, who sailed with his wife and several children.

Caliph too was built for family voyages, but it was not a replica of the *Spray*. For a start, the boat was made of steel, not wood, and its stern was round, rather than broad and square. The rig varied over time as my Dad tinkered with the design, starting as a ketch with a Bermuda mizzen and ending as a yawl with a small lugsail aft. But *Caliph* proved a comfortable and spacious home: safe, solid and reliable.

For a while, I commuted to school via dinghy then pushbike each morning. In the afternoon I called my parents on a VHS radio to collect me from the beach. When I was eleven, I left the local primary school and collected a large box of schoolbooks from the post office. My days as a student of the Correspondence School began, launching a lifetime of regulating my own workload and time. It proved the perfect apprenticeship for a life as a researcher and as a writer. I was ready to go anywhere.

But still there was more to do. Along with modifying the rig, our rebuilt vintage engine needed replacing. The internal layout was remodelled and eventually I moved from my focsle bunk with its porthole view to a spacious 1-by-2-metre cabin with a fold-out desk, bookshelves and copious cupboard space. An Alaskan filmmaker hired our boat to record white pointer sharks at Dangerous Reef in Spencer Gulf, and we spent months with the decks covered in equipment, horsemeat and whale oil, with a crew of off-season abalone divers. Our cat feasted purring on piles of bait but fled from the sharks. The white pointers circled in dark shadows around the boat, launching sudden attacks along the hull and spraying shattered teeth that glittered as they slowly sank in the water. The living conditions were rough, not great for schoolwork, so my parents sent me off to stay with my grandparents until our voyage was ready to begin.

Unlike past visits to Paris, this trip is not a holiday. This time I have coordinated my research with an old friend, Carol Harrison, a French historian from South Carolina who visits Europe regularly for her own research. We met as graduate students and have worked together before on French Pacific navigators. She is keen to help me negotiate both the archives and the landscape of old Paris.

The streets of eighteenth-century Paris bore little resemblance to the wide straight boulevards and uniform rows of Haussmann buildings in mansard bonnets that feature on Parisian postcards today. Carol tours me through the old city tucked behind the one reconstructed at the behest of Napoleon III during the Second Empire, her expert historical commentary colouring the streetscape with vanished lives and activity.

In the dark backstreets of the Marais, haphazard old buildings lean over narrow cobbled lanes. The roads twist past tiny shopfronts, and heavy wooden doors open here and there onto light-filled courtyards. Glass-fronted ground-floor apartments are topped by progressively diminishing floors culminating in tiny maids' attics. These lanes wind like tributaries down to the banks of the Seine, which once formed the major thoroughfare through the city.

In Jeanne's time, the river was crowded with barges, punts, ferries and dinghies. Livestock, horses, food and wine all flowed up and down the Seine, connecting the French hinterland to the regions and to the ships waiting in distant harbours to transport goods overseas. Not just transport though. Laundry boats also crowded the river, cleaning the city's dirty linen in its waters. The powerful shoulders of the washerwomen pounded the clothes with wooden batons, squeezed the last drops from limp sheets, seared them with flatirons and hefted washing baskets aloft.

Like the streets of the Marais, Rue des Boulangers is also old. It has followed the same narrow crooked path since 1350. Today, one end has been cut off by the busy thoroughfare of Rue Monge, constructed in the nineteenth century to improve access across Paris. The old lane is so narrow that the windows of the upper-storey apartments seem to stare at each other with stony-faced indifference. Stormwater and debris tumble heedless down the centre of the cobbled street towards a small park and market square. And beyond, the road leads directly to a side gate into the Jardin des Plantes, which surrounds the Muséum National d'Histoire Naturelle. These institutions, then the Jardin du Roi and Cabinet du Roi, were where Commerson looked for work, and where his archives and remaining collections are now stored.

Carol has rung and arranged a meeting with Catherine Crescent, who currently rents apartment 15 on the second floor of 14 Rue des Boulangers. From the outside, the building looks much as it would have when Jeanne lived here. As we wait, the large wooden door that once allowed carts to enter the courtyard miraculously lifts to disgorge a car from the depths of an underground carpark. Catherine tells us that the building was completely refurbished just a few years ago, with extra floors added. She directs us to a small shiny lift in the carpeted foyer, where soundproof doors guard access to discreet private spaces. It is hushed and modern and devoid of its history.

The rooms are small but bright: a narrow sliver of sun slides its way across the floor in the late-afternoon light. It's the only time of day the rooms get any direct sun, Catherine tells us. It's a quiet street, though, and has a friendly neighbourhood feel to it, unlike the larger busier streets nearby. Apart from the nice Italian restaurant and the humorously English-themed Baker Street Pub, most of the houses are private – concealing apartments or offices. It used to be noisy in the past, Catherine has been told, when they made metal bedframes in the street below, probably in the nineteenth century. But no matter how hard we listen, the only sound that penetrates the double glazing is the distant rumble of the Metro as it passes deep underground.

Today, these streets are filled with ghosts. The past and the present overlay one another like translucent sheets of paper until I cannot be sure which one is real and which one is not. The streets echo with the names of the scientists and writers that Commerson worked with, admired, studied and was inspired by. I walk along Place de Jussieu, named after the family of botanists who supported and encouraged Commerson and who published his work. I pass the street named after Guy de la Brosse, who founded the Jardin des Plantes. And on to Rue Linné, for the Swedish botanist who gave himself a Latin binomial, Carolus Linneaus, just as he did for every other known species on earth. His path travels at an angle into Rue Geoffroy-Saint-Hilaire, who convivially separates the Swede from his arch rival on Rue Buffon. The streets are mapped with ancient antagonisms, grand aspirations, achievements and failures.

I wonder where Jacques Labillardière lived, the botanist who wrote one of the first monographs on Australian plants and who named over 400 new species. The *savants* clustered here, in the streets around the Jardin des Plantes, a centre for research and science and knowledge, its own history recorded in patched and crumbling walls, the patchwork of buildings, and assemblage of trees and animals. I walk past the Australian wallabies in the old menagerie, which might well be descendants of those brought back on the Baudin expedition. The black cedar that Bernard de Jussieu carried back from Lebanon, in his hat after the pot broke, still stands on the path to the oldest metal structure in Paris – the gazebo of the director, Georges-Louis Leclerc de Buffon – 'the father of all thought in natural history'.

The eighteenth century was the golden age of natural history at the royal gardens, one of the oldest of all of Paris's scientific establishments. Buffon dramatically enlarged the institution, across gardens, collections, scientists and menageries. The institution hosted three professorial chairs and three demonstrators, in each of botany, chemistry and anatomy. And it was here that Commerson sought to advance his career in botany.

By 1764, Commerson was already well connected, with an impressive reputation. He had completed field trips to the Pyrenees, Provence and the Alps. Voltaire had asked him to be his secretary. Linnaeus had asked him to collect Mediterranean fish. Commerson declined the first offer and completed the second, although he broke off plans to publish his work on ichthyology to move to Paris.

Most biographies merely note that Commerson took up a friend's invitation, probably that of the influential astronomer Jérôme Lalande, to come to Paris. Lalande was said to have shared Commerson's letters with the botanist Bernard de Jussieu, who supported Commerson's move.

'The sorrow he felt at the loss of his wife and the solicitations of his friends finally determined him, in 1764, to go to Paris,' concluded one of Commerson's early biographers.

The coincidence of the declaration of Jeanne's pregnancy with their move was never mentioned.

Perhaps Jeanne taking the name of de Bonnefoi in Paris was part of a gradual transformation from peasant to servant with a hint of respectability. But the disguise did not protect them from gossip. Rumours seemed to have reached Father Beau, three hundred kilometres away. Father Beau's letter to Commerson has not been found but Commerson's impassioned reply fiercely objected to Beau's 'inquisition' of his personal behaviour and his assumption of 'a second marriage'. Instead he seems to suggest that his brother-in-law should take his share of the blame for events.

'It took you a very ardent zeal, sir, to dare to twist this rope at the expense of your own remorse: for I may ask you,' Commerson retorted. 'Who would be the cause of it if I were today, as you please to believe, on a path that were not according to God?'

The implication of a second marriage can surely only refer to Jeanne and this accusation suggests to me that it was Father Beau himself who introduced them.

'Although no-one has the right to question me in this respect,' Commerson continued, 'I can nevertheless answer your kind solicitude by assuring you that I have benefited from the changed situation . . . a truth that costs me as much to admit as no doubt it does for you to hear.'

Commerson, and his infant son, were the beneficiaries of his wife's estate. Perhaps Father Beau was concerned that marriage to Jeanne would jeopardise Archambaud's future?

But, no matter how much Commerson needed Jeanne, he did not, or could not, marry a servant, no matter how beneficial she might be.

Commerson was not merely being coy, or disingenuous, in obscuring the nature of his relationship with Jeanne. Cohabiting outside of marriage might have risked Commerson's prospects, and it certainly jeopardised Jeanne's freedom. Unsanctioned relationships were common in eighteenth-century Paris, but they carried the constant threat of a *lettre de cachet.* Such 'sealed letters' were originally a way for the king to issue orders and imprison people (a *grand cachet*), but by the eighteenth century (under Louis XIV), form letters were used to lodge accusations or complaints about immorality (a *petit cachet*). These were usually lodged by family members, to bring a wayward son, daughter or wife under control. But sometimes even neighbours or acquaintances lodged complaints. The authorities could lock up the offending parties – most of whom were women – for their own welfare.

The historian Arlette Farge records an example of a couple who objected to their son-in-law raising their grandson with a married woman. As a result of their complaint, the woman, of no relation to the complainants, was imprisoned. Perhaps it was unlikely that Jeanne's family would write a *lettre de cachet*, but Father Beau might have wished to protect the reputation of his brother-in-law, and the security of his nephew's inheritance.

There is no suggestion that Commerson and his brother-in-law fell out, however bumpy their relationship might have been. Father Beau ultimately defended his brother-in-law's legacy, and Commerson continued to write to him regularly and affectionately. He always pleaded for news of his son and signed his letters: 'I kiss you and my son a thousand times.'

He seems the very model of a devoted father.

~

Jeanne's child was due in the middle of December 1764. When the impoverished women of Paris wanted to conceal their pregnancies they gave birth at the Hôtel-Dieu hospital on the Île de la Cité, although whether it was the women who wanted to be concealed, or someone else who wanted to conceal them, is not entirely clear. They were quite literally 'confined' and not permitted to leave, nor even to have visitors. They slept four to a bed and were dressed in characteristic blue bedclothes so that they could be spotted if trying to escape.

Paris is no place to try to find a missing eighteenth-century child, either then or now. All the parish records were destroyed in fires at the time of the Commune in 1871. But Jeanne's biographer Henriette Dussourd kept looking anyway and eventually, in the Paris archives, found a child called Jean-Pierre Barret. He was 'put to the wet nurse Françoise Bernage, wife of Jean Debayé' in January 1765. No parents' names were recorded, not even that of the mother.

I walk in all directions from Rue des Boulangers, pacing the terrain Jeanne would have traversed by randomly following older narrow streets. To the west I wander by Saint-Etienne du Mont, the Pantheon and Sainte-Geneviève library, past the Sorbonne to the edge of the Luxembourg Gardens until eventually, skirting the gardens, I find myself in front of Saint-Sulpice.

The holy water fonts at the entrance to Saint-Sulpice are famous for their elegant marble sculpted by Jean-Baptiste Pigalle and adorned with crabs, fish, corals and octopus. The fonts themselves are formed from two halves of a single giant clam shell. I wonder how they got here. Giant clams – *Tridacna gigas* – are so ubiquitous as holy water fonts in France that the species and the font even share a common name – *bénitier*.

And yet these giant clams do not belong here. They are native only to the Pacific and Indian oceans.

The ones at Saint-Sulpice were a gift to Francis I (1494–1547) from the Republic of Venice, presumably originating from some East Indies trade. Giant bivalves have long embodied life-giving primordial waters with connotations of birth and feminine reproduction. Their original name, *concha*, means 'vulva'. Think of Botticelli's *Birth of Venus*. Such fertile symbolism was quickly adapted to hold Christian holy water, and the vast size of the Pacific clams – some over a metre wide – added to their mythological qualities.

The fonts in Saint-Sulpice are a ghostly shadow of their living form, as pale and bleached as the bones in the crypt. Even Pigalle's sculptures cannot revive them. There is no hint of life, of the dark quivering maw that cracks ajar like a fissure of night in coral rock. No soft velvet lips pulsing with fluorescent green and blue and yellow. No siphon drawing breath, no sudden snap, disappearing into invisibility. They are just another relic, martyred for a symbolism and a sacrifice they did not choose to make.

The fact that Jeanne's child was named at all means that she must have had him baptised. Baptism in eighteenth-century France was synonymous with naming. The rare child who was not baptised faced a nameless life, being known only by a nickname.

In Morvan tradition it would be the father's task to take the child to be baptised with the godparents, but whether he was the father or not, it seems unlikely that Commerson would have undertaken such a public activity. Her son's baptism must have fallen to Jeanne.

The 5th arrondissement was packed with churches in the eighteenth century. I initially assume that Jeanne took her baby to the current parish church for Rue des Boulangers, Saint-Etienne du Mont. On a sunny day, the interior of Saint-Etienne du Mont is flooded with light from three rows of stained-glass windows. The pale stone interior is exquisitely carved into delicate sinusoidal curves, the paired pulpit staircases spiralling like the interior of a seashell, with coralline fretwork. It is architecture that inspires heavenward.

But the parish boundaries have shifted over time. In 1764, Jeanne's parish church was actually Saint-Nicolas du Chardannet. It is a squat, heavy church with small high windows, an abundance of timber panelling, and an extreme brand of Catholic conservatism. A few elderly women wearing lace mantillas sit with bowed heads in the otherwise empty congregational seating.

Without the parish records there is no way of confirming if Jean-Pierre was baptised here. Perhaps instead, if he was born in the Hôtel-Dieu, he was baptised beneath the fading pastoral trompe l'oeil of the chapel in the foundling home next door. For it was there, at the Hôpital des Enfants-Trouvés, that Jean-Pierre was left.

The Foundling Home once stood on the forecourt of the Notre Dame Cathedral right next to the public hospital Hôtel-Dieu, which still occupies the site. The cobbled street it stood on still runs up to the door of the cathedral. Successive buildings on the Île de la Cité have been used to house the children. But the Foundling Home was always in competition with its grander neighbours, and the homes for children were progressively demolished and cleared to create a better view of the cathedral.

I had wanted to walk the street that ran past the Foundling Home, traces of it still visible in the pavement. But since the fire that engulfed the Cathedral of Notre Dame, it is impossible to get close. Barricades block off every street and bridge.

The entire forecourt is closed. At best I catch glimpses down side streets, where the sour smell of old smoke rises from cold stones. The world wept as the 850-year-old cathedral burnt, windows illuminated by the inferno as the spire erupted like a chimney of flame into the night sky. It was an ironic inversion of the fires that burnt around Notre Dame in Victor Hugo's novel that brought the cathedral such fame in the eighteenth century.

It is always sad to see something old and beautiful destroyed. We can never rebuild 'the oak forest' that made up the roof. One thousand three hundred mature oak trees, each 300–400 years old, equivalent to an old-growth forest covering twice the area of modern-day Paris, were harvested in the twelfth century to build this roof. There are not enough harvestable oaks this age left in all of Europe now. Likely neither the roof nor the forest can be recovered.

Biologists live with the grief of watching beautiful, unique and ancient things snuffed out of existence every day. Over the centuries the task of the biologist has shifted, from describing new and wondrous things, to hastily recording their all-too-sudden demise.

'To see something so beautiful and so carefully constructed be damaged by forces out of your control is very painful,' explained Jonathan Kolby, 'As a scientist who studies species that are going extinct right now, this is the feeling I grapple with more often than I'd like. The irreplaceable work of art that I worship is nature, and to watch it senselessly crumble to the ground every day hurts my heart . . . we're surrounded by burning cathedrals built across millennia and no one seems to care.'

On the day that Notre Dame burnt down, the last female Yangtze softshell turtle died with barely a murmur to note the extinction of yet another species. Countless species of molluscs vanish into extinction without anyone even noticing, before science has even had a chance to describe them. We weep for the 'offspring of a nation's effort . . . the heaps accumulated by centuries' while the world burns around us. We are weeping for the wrong cathedral.

Over a third of all children born in Paris in 1770 were abandoned. Over 1600 children were left at the doors of the Foundling Hospital in central Paris in 1772 alone. Babies could be left in a revolving babybox, or *tour d'abandon* in the outer wall. It was a system intended to keep the child safe from the elements and the identity of the mother secret.

'After pulling the rope to ring the bell and awakening the nun on guard, she runs away through the darkness with her tears and remorse,' wrote André Delrieu in his 1831 account of the facility.

How could a mother voluntarily relinquish her child to such an uncertain fate? Commerson may not have wanted any scandal, but Jeanne could have lodged a legal complaint and insisted that he marry her, or support her and her son. The historian Arlette Farge has found many such cases in the judicial records of eighteenth-century Paris, and the Commissioner frequently sided with the women's claims. Women may be hidden and silent in many official histories, but in judicial archives they defend themselves with vigour, anger and determination. They have their own opinions, beliefs and agency. I can see no reason why Jeanne would have been any different.

If Jeanne had agreed to put her son in the Foundling Home, perhaps it was because Commerson used a similar argument to the one found in the archives.

'I have been advised by a close friend, who found himself in exactly the same predicament,' wrote a young man to the mother of his newborn child, 'to do what he did and put the child in an orphanage, making sure that he could be recognized so that he could have him back when he wanted, and in fact he took him out a year after he was married. And so, dear friend, this is what I advise you to do. We can have the child whenever we want and it won't set people talking so much. Anyway, we won't be the first ones. It goes on all the time nowadays'.

There was an element of truth in this young man's claim – it was going on all the time. It was not just the poor who abandoned children they could not possibly feed. The scientist Jean-Baptiste le Rond d'Alembert was the illegitimate child of unmarried but wealthy parents who paid for his foster care. The philosopher Jean-Jacques Rousseau secretly relinquished all five infants born to his working-class lover Thérèse Levasseur between 1746 and 1752, rather duplicitous behaviour given Rousseau's philosophy of family and childhood. In 1764, the year that Jeanne's son was born, Rousseau was publicly exposed by Voltaire.

It is true too, that parents could, and did, reclaim their children. While many children were abandoned without identification, others had detailed notes attached, with names, addresses, tokens, buttons, coins, medals, carvings, cloth or ribbon so that they could be reunited with their parents. There is a poignant display of such tokens, from children who were never reclaimed, at the London Foundling Hospital: testimony to so many parents' lost hopes.

I don't know if Jeanne left a token or if Jean-Pierre Barret was her son. But this name suggests a label by which he could be identified. It is not only her own surname (as she spelt it), but also the names of her father and brother. This child's name implies familial bonds of affection. The name 'Jean-Pierre' connects the mother to her son and the daughter to her father and brother. She was neither alienated from her own family, nor from her child. Perhaps she planned to return in more auspicious times to retrieve him.

Retrieving a child was not always successful though. Farge also recounts the story of a mother who asked for her twelve-month-old son to be brought back from the wet nurse to Paris. But the child died on the water cart he returned on. The distraught mother recognised him only from the linen of his layette. She had not seen him since the day he was born. Babies were such fragile little creatures, so prone to dying, in orphanages as much as in the care of their own mothers. More than half of the children left in the Foundling Home did not survive their first year.

If Jeanne had intended to retrieve her son, she never got the chance. In March 1765, within just a few months of being given to the foster nurse, Jean-Pierre Barret had died.

It is difficult to imagine Jeanne's role in Commerson's scientific work when all the names I know from the Jardin du Roi were men. I know a few wives and daughters who helped with their work, but mostly these women are silent, invisible and almost unknown to us. And yet, in truth, the women of eighteenth-century France were both vocal and conspicuous. We can see them in the plates of Diderot's *Encyclopédie*: in the workshops, on the streets, in the kitchens, in the country, writing, printing, riding, labouring, sewing and making buttons. There could be no doubting the value of women in skilled trades after engineers had to take instruction in Roman hydraulics from peasant women in Pyrenees bath towns to build the 240-kilometre-long Canal Royal en Languedoc (now the Canal du Midi) in southern France.

Women were far from silent in the sciences, despite being denied education. The brilliant physicist Émilie du Châtelet was famous in Jeanne's time, and her publications and translations inspired generations. Excluded from male-only intellectual gatherings, she had dressed as a man to gain entry. At the time Jeanne was in Paris, an impressively experimental book on putrefaction was published anonymously to critical acclaim, although few would have guessed that its author was Marie-Geneviève-Charlotte Thiroux d'Arconville.

Jeanne would certainly have known of, if not met, Nicole-Reine Lepaute, the daughter of a valet and self-taught mathematical genius who calculated the trajectory of Halley's comet, the transit of Venus and a solar eclipse, and compiled star guides used by astronomers and navigators. Lepaute collaborated with Commerson's good friend Lalande and Commerson counted Lepaute as one of his friends and advisers.

Nor were women absent from the Jardin du Roi which, unlike the universities, allowed women to attend classes alongside men. A commitment to public education meant the classes were free, open to all and in French (rather than Latin). Commerson must already have been training Jeanne to help him in his botanical work. She would surely have had cause to study, admire and learn from the impressive artwork and dissections of Françoise Basseporte, the garden's royal botanical painter. Even the most senior figures like Michel Adanson and Linnaeus deferred to Basseporte's expertise in distinguishing species. Jeanne might even have attended the anatomy lectures of Marie Marguerite Bihéron whose exquisitely lifelike wax models were an international sensation. Such study might have been a good distraction.

There were plenty of role models for Jeanne in these circles. From all walks of life, these daughters of nobility, farmers, apothecaries, servants and tradesmen were linked only by their talent and intellect, and their determination to pursue their interests and passions no matter what obstacles were put in their way.

In late 1766, Commerson was offered the position of Doctor-Botanist and Naturalist to the King on board an expedition around the world, commanded by Louis Antoine de Bougainville, an important post on the first truly scientific expedition of exploration. Bougainville's voyage was the beginning of a program of state-sponsored scientific discovery which would yield an unparalleled haul of new knowledge.

'It is one of those events which is an epoch in the political and literary world,' said Commerson.

The science on Bougainville's ship was, like all of the French expeditions that followed, generously funded. Commerson's salary of 2000 livres (less than the expedition commander's, but more than either of the captain's of each ship) and an equipment budget of 12,000 livres signals the beginning of a great campaign of intellectual discovery – the birth of anthropological, geological and biological sciences in the southern hemisphere alongside the traditional naval disciplines of astronomy, celestial navigation and cartography. Through all the political upheavals to come, France maintained a national commitment to well-funded scientific voyages – through kingdoms, republics, empires and restorations.

Commerson's instructions were to make observations and collections on the coasts and interiors, to keep precise records so that he could give an accurate account on his return. In order to do so, he was to be provided with 'all the items and belongings that will be necessary for your observations' with the assistance of Pierre-Isaac Poissonnier, medical consultant to the King, inventor of the ship's desalination unit and an influential ally.

Commerson was thrilled. Even as he outlined his dilemma as to whether to go in a letter to his brother-in-law, it was clear that his mind was made up. Should he stay in this life of mediocrity to which he had happily resigned himself? Should he stay for the ties of his son and family? Or should he travel in the footsteps of Vespucci and Columbus, and take this singular opportunity to witness all manner of new and wondrous things?

His friends all urged him to go – Bernard de Jussieu and Poissonnier as well as Auguste Charles de Flahaut de la Billarderie. The man he called '*mon intime*', Clériade Vachier, from medical school and now practising in Paris, supported his decision, as did Nicole-Reine Lepaute.

By January, Commerson had made up his mind. 'They tell me when I come back the Order of St Michael will be mine, posts and pensions will become available. Every door will be opened to me,' he said.

But Commerson would need someone to assist him with his collections. His health was not the best. Everyone knew his limitations. And even this was generously accommodated.

'I have been given a valet, paid and fed by the King,' he declared happily.

Who would be the best person to take? Where could he find a servant with the knowledge, skills and expertise to help him? And, more importantly, the extraordinary level of stoic endurance he would require?

The official plan was, it seems, for Jeanne to stay and look after the house. Just before he left Paris, on 14 December 1766, Commerson wrote a long and detailed will, mentioning Jeanne in the eighth clause.

'I bequeath to Jeanne Barret, known as de Bonnefoi, my housekeeper, the sum of six hundred livres,' he wrote, 'paid in one amount and this without prejudice to the wages I owe her from the sixth of September 1764 at the rate of one hundred livres a year, stating furthermore, that all the bed and table linen, all the women's dresses and clothes I may have in my apartment are her personal property, as well as all the other furniture, such as beds, chairs, armchairs, tables, chests of drawers, excepting only the above-mentioned herbaria and books and my personal effects left to my brother as stated above. I wish that the said furniture shall be handed over to her without hindrance after my death, even if she retains for one year following the present one the apartment I will be occupying at the time, even if this were only to give time to sort out the collections of natural history that are to be sent to the Royal Cabinet.'

This document is the only time that Commerson admitted that he knew Jeanne before the voyage and that she had worked for him in Paris. It is a telling revelation. If he had truly wanted to keep her presence a secret, he could have entrusted the matter of reimbursement to his dear friend Vachier. But Commerson does not leave Jeanne's entitlements to chance. Even at the risk of exposure, he puts her interests first.

5

A NEW COMMISSION

Nantes, 1766

SURELY ALL SAILORS LONG for undiscovered lands and unspoilt wilderness, to press their foot into sand that has seen no human imprint before. Is that not what we are all searching for? There is something profoundly thrilling about stepping onto untouched shores. How could anyone resist this call of the wild? Even when we know that our very presence – the act of observation – destroys the thing we were most desirous to find?

It is what I loved most, what I miss the most, from my childhood, that notion of stepping on land that seems untouched by human hands. It was a fantasy easier to maintain in childhood, when I could not see the human impact so well, when there was less pollution, fewer jets, and when the world's population was half what it is now.

I suppose our early days of sailing around the islands and bays of Eyre Peninsula predisposed me to expect that the world was a place of emptiness and wilderness where people were few and far between. My notion of the world was one of stark limey cliffs rising out of wild blue seas, their flat ridges blanketed by a mosaic of mossy mounded coastal vegetation in a patchwork of grey-blues and olive-greens. Swathes of grass shimmered and rippled like molten gold across the inland, flowing around silvered salt lakes. The white sandy beaches vanished into the distance, backed by long windswept dunes held in place by razor-sharp grass. The sand was so fine and white that it packed squeaky hard as icing sugar beneath your feet. And I pranced like a windswept waif across the fragile surface, fair hair flying, leaving no mortal trace on the hard surface while my heavy-footed parents sank into the softness beneath.

Boats offer a particular advantage to maintaining this sense of isolation. Should some inconvenient fishing boat or picnicking party happen to intrude on your own private wilderness, playing loud music, or launching jet skis or riding trail bikes, you can haul up anchor and move somewhere else. Good fences may make good neighbours, but that is nothing compared to a few good sea miles. With their endless capacity to put distance between yourself and whatever is bothering you, boats offer the ultimate form of escape from the daily irritations of human interaction.

Despite the obvious attractions, I still wonder why people signed up for these early voyages. The risks were so high. A great many did not survive. But perhaps it is the same sense of restlessness that drives so many to emigrate, to move, to start a new life in a new land. For some, like Commerson and Bougainville, the glory and the rewards were clear enough, both financial and personal.

For others, like the sailors, the risks of a life at sea were perhaps no worse than those on land, and the benefits of a set naval income (which could not readily be spent) worthwhile. Yet Jeanne gave up a secure and safe position looking after Commerson's Paris apartment while he was away, answerable to no-one, to embark on an unknown adventure in a world where women were distinctly unwelcome.

What causes a particular group of people to set off from their homes with no clear certainty that they will ever return? I once traced back my family tree, over four or five generations, stopping only when I reached the adults who were not born in Australia. I identified at least 34 ancestors who had emigrated to Australia. What distinguishes these people from the ones who stayed behind – the ones who never seemed to go anywhere?

I read about a genetic study done on a 3000-year-old body pulled from a bog somewhere in Cheshire. The local butcher in a nearby small village turned out to be a direct descendant. Folks in Cheshire knew their place it seems. They were not migrants, sailors or nomads.

There must be something inherited in rootlessness. People descended from colonists – the non-indigenous inhabitants of Canada, Australia, the United States and New Zealand – move house, move state, move long distances within their own countries far more frequently and far further than residents of the United Kingdom and Europe.

It's possible that some of my ancestors fled religious or cultural persecution: German Lutherans from Prussia or Scottish highlanders from the clearances. Others might have been escaping poverty and hard times: Cornish miners in search of rocks to quarry. But many left behind perfectly respectable businesses: the champagne-makers, hoteliers, farmers and teachers trying their luck in a new world. What made them leave and others stay? Surely it can't just have been the weather.

And what about the explorers? They were in search of something different. They were not looking for a new home, but were, by and large, not afraid to leave their old one. Many of them would not return. They must have known the risks. But then, they were mostly young, in their teens and twenties, when their perception of risk, their own sense of immortality, was notorious.

Commerson knew the risks. Before he left Paris, he worked on compiling a martyrology of botany, so great were the numbers of botanists who had not returned from their travels. But they still did it, for the lure of fame, fortune or family tradition.

Jeanne was not seeking fame. She was not following a family tradition. I don't think she was coerced, nor excessively devoted. I don't think the money would have been any better than if she had stayed in Paris.

What if Jeanne sailed simply for the adventure? For the opportunities that might arise in a new and unexplored world? Commerson was passionately enthusiastic about the trip. Perhaps she shared that excitement. Her bravery, stoicism and self-determination may well have been one of the reasons Commerson was prepared to risk so much to retain her company – the ire of his in-laws and of polite society, and the consequences of defying naval ordinances in a position that could be the making, or breaking, of his career.

Perhaps Jeanne did it for exactly the reason so many others chose to go to sea, as Herman Melville wrote, because they were 'tormented with an everlasting itch for things remote . . . to sail forbidden seas, and land on barbarous coasts'. She told Bougainville that she embarked simply because the notion of sailing around the world had 'piqued her interest'.

That seems as good a reason as any. We should take her at her word.

~

I wonder what stories stirred Jeanne to travel. When Jean-François de Galaup de Lapérouse set sail from France in 1785 on his famously ill-fated voyage, the only novel listed in his extensive shipboard library was *Robinson Crusoe*. Like many other sailors, Matthew Flinders was famously 'induced to go to sea against the wishes of my friends' from reading the same book.

Jeanne was not likely to have been inspired by a character in a book she could not read. The philosophical novels of Voltaire and others would, presumably, have passed her by unnoticed. Perhaps some travelling theatres passed through La Comelle to entertain with Molière or something colourful and Italian? What message could she take from Charles Perrault's classic fairytales with long-suffering women and opportunistic men? Or perhaps she preferred the heroines in Madame d'Aulnoy's fairytales, who were rewarded for overcoming less morally accountable challenges. It may have been the songs of the troubadours, the epic battles of Charlemagne, the stories of French saints or simply the fables of Tybalt the Cat or Hirsent the She-wolf that lit Jeanne's eyes as she sat around the fireplace. Or perhaps, even in Burgundy, they told the legendary stories of Jeanne d'Arc, the Maid of Orleans, and the naval triumphs of Jeanne de Belleville, the Lioness of Brittany.

She could not have known that one day it would be Jeanne Barret who featured in such stories.

By the 1760s the general shape and structure of the world had been mapped by the European cartographers. Africa, Europe and Asia all had their characteristic shapes. Most of South and North America were known, save some vagaries around Canada and Greenland. But the western Pacific was still largely uncharted. Both Australia (New Holland) and New Zealand were pencilled in, but much of their coastline was indicated only by the vaguest of dotted lines. They were barely recognisable. The detailed maps of South-East Asia petered out beyond New Guinea into a smattering of islands that bore little resemblance to any modern geography. There was a lot of guesswork for the Pacific, at least for Europeans.

A traditional account of history would mention the domination of the Pacific by Spanish, Portuguese and Dutch trading powers. Their focus was on Asia, including the Indian subcontinent and South-East Asia. Australia and the oceanic Pacific, with poor opportunities for trade, were left largely unexplored by these powers. As the influence of Spain and Portugal declined, however, the European battle for imperial control shifted to France and Britain. While France emerged as one of the leading powers in Europe, the much smaller Britain positioned itself in the battle for control over global trade, quite probably on the back of its well-developed skills in piracy. It is a well-rehearsed story told over centuries as if it is the only version of the truth.

But there are other ways to tell this story, other truths. The Pacific was not unknown or even unmapped. The region had been comprehensively explored, discovered, navigated and settled for thousands of years before Europeans first ventured into its waters. The waves of exploration and migration, settlement and colonisation, war and trade across the islands of Asia and Australia and around the Indian Ocean are as complex and ancient as any Eurocentric human history.

The greatest maritime exploration of the Pacific belonged to the Polynesian people, who swept across Samoa, the Society Islands, New Zealand, Hawai'i and Rapa Nui in an immense expansion between 800 BCE and 1290 CE. In the course of 2000 years, from the founding of the Greek states to the end of the Crusades to Jerusalem, Polynesian people expanded across the largest ocean on the planet, settling islands scattered over some 35 million square kilometres with a single coherent cultural group sharing language, customs, myths and skills.

Pacific history does not begin with Europeans. As Tongan anthropologist Epeli Hau'ofa tells us, Pacific Indigenous history is not a footnote to the history of empires. It is, in fact, an account of its own great maritime empire, whose navigational and shipbuilding achievements predated and exceeded those of the Vikings, the Portuguese, Spanish and Dutch by centuries, if not millennia.

The narrative of history is, however, that of the written word. And with one swift definition, all oral history is disowned, truncated, belittled and silenced. This 'unwritten' history, before European contact, becomes 'prehistory' – relegated to folklore, mythology and superstition.

Even when Europeans encountered these great navigators, puzzled over the mystery of their presence on such isolated islands, admired their boats and love of the sea, it seemed that no-one bothered to ask them what they knew or how they travelled. And those who sailed with Bougainville into the Pacific would not be much different. It was as if it was beyond their imagination that people could travel so far, across such vast oceans, in such small vessels.

Louis Antoine de Bougainville was a veteran of the Canadian wars, a talented young man of modest origins fortuitously supported by well-connected family friends, who had the ear of the influential Jeanne Antoinette Poisson, chief mistress to Louis XV and better known as Madame de Pompadour. He seems to have been one of those clever capable men who did well at whatever he attempted and who was, by and large, admired by those he worked with. Mathematician, diplomat, ethnographer, naval commander, humanitarian, strategist and colonist.

Dismayed by the loss of the Canadian colonies to Britain, Bougainville hatched a plan with his good friend, the sailor and privateer Nicolas-Pierre Duclos-Guyot, to set up a new colony on the unpopulated islands off the South American coast, which he named the Malouines after the port of St Malo but which are now known as the Falklands.

The French crown was willing to support the plan, but not to finance it, so Bougainville drew on his considerable personal connections to fund the building of two ships as an expedition fleet to the South American islands. The second ship, the *Étoile*, which Jeanne and Commerson would eventually join, was captained by François Chenard de la Giraudais, another veteran of the Canadian wars. Between them, these friends made several trips to South America, to the new colony on the Falklands, as well as to the Strait of Magellan and to the Spanish settlement of Montevideo, where their presence raised the suspicions of the Spanish authorities.

Spain was not happy with a French settlement appearing so close to their southern colonies. And so, while the colonists were settling in to their new home, the Spanish court was protesting to its counterparts in France. Madame de Pompadour had died. No-one knew whose influence would rise or fall in the circle of royal favourites. France could not afford Spanish animosity. It relinquished its claim to the Falklands and Bougainville's nascent colony suddenly found itself without a home.

Bougainville's orders were to sail the frigate *Boudeuse* to the French settlement on the Falkland Islands, and formally return the islands to the Spanish. If Bougainville doubted the Spaniards' capacity to retain their newly acquired asset from the grasp of the English, there was nothing he could do about it. His diplomatic duty completed, he would be joined there by the store ship *Étoile* and the two vessels would continue on their voyage around the world.

The expedition would leave from Nantes, where the *Boudeuse* was being purpose-built for the voyage by Jean-Hyacinthe Raffeau at the Indret shipyard on the Loire River. We can see the busy riverside of Nantes, including the shipyards, in a series of nineteenth-century watercolour sketches by William Turner. Upright terrace houses on land mirror the forest of masts on water. Figures seem in constant motion on boats, in carts, on foot; buying, selling, building, loading. And low arched bridges and imposing chateaus rise above the markets. You can almost smell the eels sizzling in their butter.

As Jules Verne described his home town, Nantes was dominated by 'the maritime movement of a big city of commerce, the point of departure and arrival for many long-haul voyages. I see again the Loire, its multiple arms connected by a league of bridges, its quays encumbered with cargoes under the shade of great elms . . . Ships docked in two or three rows. Others go up or down the river. No steamboats . . . we had only the heavy sailing ships of the merchant navy. What memories they stir! In imagination, I climbed their shrouds, I hoisted myself into the crow's nest, I clung to the apple of their masts! What a desire I had to cross the quivering board which connected them to the quay and set foot on their deck!'

The Loire River was the departure point for ships and voyages, both real and imagined, for centuries.

I want to find where those eels end up, the ones that grew up in the rivulets and creeks of the Morvan before heading down the Loire River. Not all of them make it to the sea. Many of them are caught in nets on their way down the rivers, and cooked for dinner. But eels are hard to come by nowadays, even in Nantes.

I have joined a friend who lectures in French at Flinders University for a tour of historic Bretagne, the larger region that once comprised Brittany. Christèle's family live in France. Her sister lives in Nantes, which was once Brittany's capital until a departmental redistribution placed it outside its historic region. This port on the Loire River is too important to miss. It was not just the launching point of Bougainville's expedition and the birthplace of Jules Verne's fictional voyages but also the departure point for the long migration of the Arroux River eels to the Sargasso Sea.

There is a worldwide shortage of eels, from Japan to France, and they command a high price for their scarcity. Christèle's sister tells us you can still get them, fried in butter and parsley with fresh new potatoes, in a little restaurant on the south bank of the Loire, aptly named La Civelle. It is a short ferry ride from the city centre down the river and past the island. The cross-hatched beams of the old docks line the river, broken here and there by industrial inlets – slipways, drydocks and berths. The skyline rises with old warehouses and sheds. Cranes lean over them, no longer a sign of maritime activity, but rather a new phase of real estate development, repopulating the industrial landscape with music, art, shops and housing.

As we head down the river we pass the steampunk theme park inspired by Verne's work. A giant elephant, two storeys high, stalks between the old warehouses, carrying excited passengers on its exotic umbrella-ed panniers. A group of school girls squeal as the elephant swings its steaming trunk towards them. The workers who once trained in the shipyards and laboured on the docks have now brought the skills that built ships and engines to the production of elaborate carousels, street dragons, spiders and birds. The craftsmanship of centuries merges with the creative imaginings of art to give aesthetic life to new creations in an unexpected modern industry.

Our ferry docks at an enclave of cheap colourful restaurants popular with local office workers and visitors alike. We take a table on a dock looking out over a cluster of chairs that tumble down a cobbled boat ramp towards the river. We relax in the sun with a glass of crisp white, as we wait for the eels, which arrive freshly cooked and sizzling hot in an old pan, as simple and unadorned as they have ever been cooked along this river.

As we finish up our lunch, I watch a small recreational fishing boat slide down the river past the empty decaying dockyards of the once hectic commercial port. And it occurs to me that this is the only boat, other than the ferry, we have seen in the last two hours along this stretch of the Loire, once one of the busiest ports in all of France.

It was unusual for navies to provide a new ship for a voyage of exploration. Normally they would offer tired old ships, already close to the end of their short sailing lives. If they survived the long journey they were typically retired to serve as brigs and prison hulks, as storage or broken up for firewood. Appearances do not matter much for voyages of discovery. Any European ship, they might have argued, was enough to impress an islander in a pirogue. A cannon and a few muskets enough to subdue with superior firepower. A handful of trinkets enough to gain friendship and trade. These were not military voyages. They did not plan for violence or invasion, even though this seemed to follow inevitably in their wake.

And yet Bougainville received a brand-new ship, a fifth-rate frigate no less, for his journey, adapted with French innovation to carry the extra weight of 26 12-pound, rather than the usual 8-pound, cannons on the upper gun deck. The *Boudeuse* also had six 6-pounders on the quarterdeck and focsle. The ship could carry a full battle complement of 252 men, including twelve officers, into war, although fewer crew were required for peacetime activities.

Bougainville's second-in-command, Duclos-Guyot, oversaw the fit-out. The ship was 40 metres along the gundeck, 10.5 metres across the beam. The bluff bow was fitted with an ornate figurehead. The sketched plans show a wild-eyed, curly-maned lion astride the head timbers, paws clamped over the royal crest. The wide square stern was studded with a single row of paned windows above two hinged ports for stowing spare timbers and spars. The tafferel above was tastefully enclosed in ornate neoclassical carvings, framing another royal crest, which sat among an array of standards and flags. The carvings would have been painted golden yellow, along with the trim on the upper timbers of the topsides, setting off the bright red hull and tarred black underbelly. Once complete, the ship was decorative but restrained – modest compared to the gilded glories of earlier periods, more suited to the refined tastes of the age of Enlightenment, yet indisputably signalling prosperity and impeccable style.

Why did Bougainville's expedition warrant a brand-new warship? Was it a compliment to the commander, a war hero and personal favourite of Madame Pompadour? It probably had more to do with the Spanish and the necessary pomp and ceremony for handing over the Falklands to his Catholic Majesty without losing face. Fellow Europeans require much fancier trinkets to impress than those required for the people of other lands.

Despite the frigate's military nature, the demands of the expedition came first. The central upper deck might have been nominally dedicated to artillery, but beneath that a false deck provided space for the crew, shared with cows, sheep and chickens. Towards the stern was the gunroom, and two small cabins of 2 by 2.5 metres each for the chaplain and the surgeon. Behind this, Bougainville's cabin stretched 6 metres to the row of windows across the stern and occupied the full 7-metre width of the ship. Two small officer's cabins squeezed to the front of this space on each side, while the remaining officers' quarters were sheltered under the quarterdeck above.

It was, all in all, a grand ship to look at and impressive to sail, even if it was not ideally suited to an expedition. It drew too much water for close coastal work and could not store enough supplies to feed its crew for a long journey.

Bougainville, unusually, appointed his officers on the basis of experience rather than their connections. As captain of the *Boudeuse*, he appointed Nicolas-Pierre Duclos-Guyot, an 'officer of the blue': a Breton captain who earnt his position through merit rather than through nobility. Naval positions were more often given to influential aristocratic 'officers of the red'.

The *Étoile*, too, was under the command of a Breton sailor. Bougainville had sailed this ship to South America before. He knew it was a reliable and solid vessel, as was the captain, François Chenard de la Giraudais. At 40, La Giraudais was one of the oldest and most experienced men on the expedition. He had been at sea, with his father, since the age of five, slipping between the merchant and royal navies, successively commanding larger fleets and bigger battles. Valued in war, like most officers of the blue he was unceremoniously dumped by the navy in peacetime. He knew enough not to rely on the navy to make his fortune. He was a capable and competent sailor, generous and easygoing, if sometimes too cautious for his less experienced expedition commander.

The other officers of the *Étoile* were also pragmatic and down-to-earth. Jean-Louis Caro and Joseph Donat seemed to have had little interest in personal intrigues or philosophical debate. They were Breton sailors, plain and simple – loyal, stoic and excellent at their job. They set their own courses.

Not to say the expedition didn't have any influential aristocrats. It was the French navy after all. Charles-Henri-Nicolas-Othon d'Orange et de Nassau-Siegen, most often known as the Prince of Nassau, was king without a country, perhaps even without claim to a title. Bougainville had met him at a dinner given by the Comte de Maurepas, a prominent statesman. Nassau-Siegen was a charming and erudite young man, with naval experience, but also a prodigious debt and troubled reputation. Bougainville invited him to join them as a volunteer, but over the course of the journey he increasingly took on the duties and responsibilities of an officer.

Bougainville departed from Nantes on 15 November 1766 while the *Étoile* was still being fitted out in Rochefort. Just two days out to sea the *Boudeuse* was struck by a violent squall that snapped the fore-topmast, then the main-topmast, which carried away the head of the mainmast. Bougainville had no choice but to return to Brest for repairs.

It takes time, and long sea trials, to finesse the balance of a sailing boat. *Caliph*'s rig took years to evolve, as did the distribution of ballast. We collected our first ballast from the northernmost of the Bicker Islands off Port Lincoln. On a calm day, we nudged *Caliph* in against the deep-shelving shoreline to load perfectly rounded football-sized boulders directly into the bilge, just as generations of coastal trading ketches had done before us.

At the height of the age of sail, ships were fragile constructions of timber and rope that floated with delicately balanced precision between air and water – whether riding high and light, or fully laden and wallowing, whether in shallow waters with a short chop, or in great rolling swells like mountain ranges, whether running fully rigged before the wind, or beating mean and tight into it.

Like many French ships, the *Boudeuse* pushed the boundaries of innovation in naval architecture. It had been built with an 'enormous' tumblehome, the upper deck of the hull curving inwards like a tulip-shaped wine glass. Tumblehome improves stability, lowering the weight carried in the ship. It makes the ship more difficult to board at close-quarters, while the sloping flanks deflect cannon balls more effectively than straight sides. But the narrowed width also reduces the angle between the base of the shrouds and the top of the mast, essentially making it harder to support the tall rig. On the *Boudeuse*, this problem was compounded by the heavy load of stores required for the long voyage planned, which were clustered at the bases of the masts and along the keel, making the masts vulnerable when the ship rolled.

This is why new ships require running in. It takes time to get to know their individual habits and temperament. It takes time to calibrate the distribution of ballast against the resistance of the water and the windage in rigging and sail, in order to achieve perfect performance. New ships need trimming in sea trials to ensure that the rig and the hull are balanced and stable, ballast shifted, rigs modified, masts and spars extended or shortened. It would be disingenuous to expect a newly built ship to embark on a long sea voyage without the benefit of adequate sea trials but there was no time to run in the *Boudeuse*.

Bougainville ordered the masts shortened, the upper timbers re-caulked, and replaced the 12-pounders with 8-pounders, substantially reducing the ship's weight. But he still had his doubts about its seaworthiness for a long voyage. He considered sending the ship back from the Falklands and continuing in the solid and reliable *Étoile*, although in time he would appreciate *Boudeuse*'s speed and rail over *Étoile*'s lack of pace.

Those on board would be equally tested – both old hands who could be relied upon and new ones who had yet to prove their worth. They would have to be as tough as the brown burrowing bivalves, the sea dates, that buried themselves almost invisibly within the rocks and withstood all the pounding ferocity of the waves.

'Glory requires, like Fortune, a race of men both hardy and tenacious,' Commerson would say as he prepared to embark.

Of course, it was not just the men who would have to be hardy and tenacious on this particular voyage, but a woman too.

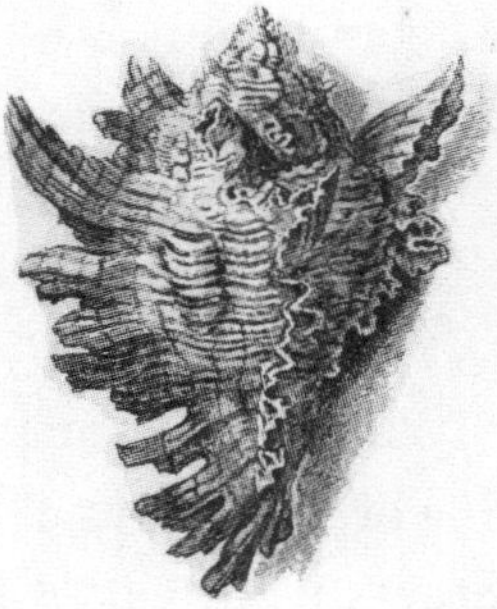

6

FITTING OUT FOR THE VOYAGE

Rochefort, 1767

IT IS HARD TO imagine the forest of sailing ships that once crowded the seaports of the Atlantic, Pacific and Indian oceans. It is hard enough to even imagine what working eighteenth-century ships might look like, let alone how they sailed and how those who sailed on them lived and worked.

The world of these old ships has completely vanished. Long-dead forest timbers were never intended for a life at sea. They mouldered with rot and damp, were tunnelled by teredo worms, chewed and gnawed by anything with teeth, beak, shell or drill. They buckled, twisted and splintered under the constant strain of wind and water against warp and weft. Few timber sailing ships have survived from before iron frames gave their spans rigidity.

Only two ships remain afloat from the mid-1700s: the English *Victory*, saved by its fame as Nelson's deathbed; and the salvaged shipwreck of the American gunboat *Philadelphia*. The oldest ship still found regularly sailing offshore is the iron-hulled windjammer the *Star of India*, built almost a century later in 1863.

In every harbour I visit, I look for traditional boats, trying to recapture something of that lost world depicted in vast detail in Claude-Joseph Vernet's paintings of French ports. The port of Saint-Malo seems a promising place to look, during a lunch break from a writing festival several years ago. A light drizzle keeps tourists at bay but the harbour itself is crowded with vessels of all shapes and sizes. Large ships loom in distant mist, grey and streaked with rust, overhung by industrial-scale cranes. The fishing fleet bristles with aerials and receivers, sonars and radars, back decks awash with machinery for hauling monster nets and pots. Closer by, a motley collection of recreational vessels fills every available space: small and large, sailing yachts and motor cruisers. Most gleam white and silver with glass fibre and aluminium, ferro-cement and steel. Clean, sharp, modern and uniform.

But my eye is always drawn to the irregularities of timber and rope, the dark spars held aloft with deadeyes and lanyards, the smoothed anomalies of timber planking. The 'trad-riggers' are not hard to spot: dark among light, complexity among simplicity, vertical among horizontal, heft among the insubstantial. They stand out, like old-growth trees in a plantation, against a mono-cultured forest of silver 'radio masts' that give modern sailing boats their Marconi rig nickname. Once this forest would have been a veritable ecosystem of ocean-going, coastal and riverine vessels, all grown from oak, hemp and cotton, all built for war, trade and survival, and all curved into exquisite shapes by the pressures of the water and wind that powered them.

As it happens, the *Étoile du Roy* is in Saint-Malo port and open to visitors. This vessel is a replica sixth-rate frigate, built in Turkey in 1996 to star as the *Indefatigable* in the TV series *Hornblower*, then as Nelson's *Victory* in the Trafalgar commemorations before transferring to a French flag in 2010. The *Étoile du Roy* is a little larger than the *Boudeuse* and larger again, though not as roomy, than the *Étoile*. An expert eye could see it is not a French-built ship, with less ornate head rails that terminate further forward. The enthusiast would complain that modern construction constraints have compromised sailing ability, but the reconstruction is as close as I can get to the ship Jeanne sailed on.

I have the *Étoile du Roy* to myself when I visit. The bright yellow and blue paintwork gleams against the black hull and the rain slicks the decking timbers shiny and treacherous.

'One hand for the ship, one hand for yourself.'

I run my hand down the timber handrail, remembering the ease with which I once swung myself around the confines of a boat, the monkey-like flexibility of childhood in an environment that required agility, balance, strength and constant vigilance. I duck my head, grip tight to the handrails and step, slowly and very carefully, into the dark.

Only the upper decks are open to the public. I suppose most visitors don't find the holds and storerooms and bilges very interesting. I like seeing the bones of the ship, the ribs and knees that give it shape, the long keel that holds the ship together, the ballast that weights its stability. The lower decks expose the ship's shape and condition – the split seams, the hogged keel, the cracked ribs. The patches and repairs reveal the constant struggle against weather, worms and waves. I resist the urge to dig my thumbnail into soft timbers to check for rot, the maritime version of kicking the tyres – a legacy of a childhood spent in too many boatyards.

Peering down a companionway leading to the lower decks, I glimpse an engine room. There are limits to authenticity on any ship. I will never find a ship with the original bilge pumps in working order, or an operational original woodstove in the galley, nor a revolutionary cucurbit, designed by Poissonnier, to extract fresh water from salt. No-one wants their ship to sink or burn to the ground. No modern ship will have a crew of over 200 men, living, breathing, sleeping, belching and farting in close quarters. No fetid water in swollen casks, no weevilled biscuits or salt-encrusted meat, no lack of fresh vegetables. No livestock loaded in crates on the deck: goats, cattle, horses and chickens. No scurvy, no fever, no wasting diseases. No-one wants their crew half dead either.

I try to fill each empty space with people and equipment. Mentally, I stack all the spaces with stores and livestock. I fill the decks with small boats and cages, temporary cabins for the galleys. I remind myself to triple the size of the mooring lines that coiled, in those days, like giant boa constrictors. I take photos, too dark to see properly, too cropped to capture the shape or size of the spaces, knowing that later I will want to see the thing that is just out of view. I try to remind myself that the past is not just 'us in funny clothes', but the effort required to sustain the reconstruction is too hard. I can barely even see this ship as it would have been, let alone 100 ships packed into a busy port crowded with people shouting, ordering, working, loading, carting, cleaning, building and repairing.

So how am I to find a single young woman, some 250 years ago, dressed as a man, with her head down to avoid attention, on her way to join a ship that was to sail around the world? Much less find her place in a world where women were not wanted?

~

Commerson left Paris directly from Poissonnier's house but I think he travelled alone. He does not mention a servant or a travelling companion in a letter he wrote to his brother-in-law. He uses the singular pronoun in his letter, not a plural.

He took just three days to arrive in Rochefort from Paris, a cracking pace for a horse-drawn diligence in the middle of a freezing winter. We know exactly what route he took, the conversations he had, and about the accident in Niort where he was nearly crushed by a carriage because of 'the clumsiness of a half-drunk postilion', reopening an old wound that would plague him for the rest of his years. But we don't know where Jeanne was.

To all intents and purposes, she was still in Paris, looking after the house. This was the story Commerson told in his will. Likely this was the story they told everyone else. But Jeanne de Bonnefoi, the housekeeper, disappeared from Paris without a trace.

In her place, a quietly confident young man, Jean Barret, former valet to a Genevan gentleman, appeared in Rochefort, looking for a new position.

It was not a good thing for a woman to dress like a man. Moses was quite adamant about the matter:

'A woman shall not wear a man's clothes, and a man will not put on a woman's garment: anyone who does this is an abomination to God.'

Eventually the priests realised that perhaps there might be circumstances where a woman could dress as a man without sin. It was better to wear men's clothing than none at all, obviously. By the time of the *ancien régime* in France, the avoidance of death or rape was deemed a sufficient excuse for committing the mortal sin of wearing 'the wrong trousers'. But otherwise, such 'cross-dressing' remained a crime.

Women were arrested for being 'disguised' as men. A chambermaid, Suzanne Goujon, was sent to prison for dressing as a man and renting lodgings 'with no female clothing in her possession' in 1749. So was Marguerite Goffier, who went to the cabaret dressed as a man. But still they persisted. Françoise Fidèle dressed as a man to enlist in the Paris militia. She indignantly declared that 'she was a good girl and had done nothing to offend her honour and that she was not the first in her family to disguise herself in order to enter the service'.

It was not just French women, either. English women too, like Hannah Snell, Mary Lacy and Mary Anne Talbot, signed up and served as men in the British navy in the late eighteenth century and later became quite famous for their exploits. Or 'William Brown', the African-American woman who served as no less than captain of the foretop on the *Queen Charlotte* for eleven years, continuing even after her gender was revealed. Maybe women were not supposed to be on ships, but they were not uncommon either.

Successfully passing as a man takes more than just clothes. It is also about behaviour and attitude. Men are louder and they talk more at work. They fart, belch, sneeze and piss louder than women. They take up more space: they sprawl, stride, stretch and sit with legs spread, feet apart, arms open, chest exposed in challenge. To sit like a woman, legs to one side, coiled, crossed and covered, is to adopt a defensive posture. In order to blend in, Jeanne had to 'stand out' like a man.

The English women frequently mentioned having to stand up to men, shout them down and show no fear. Mary Lacy had no choice but to fight one young man who challenged her. She was not as strong as him, so she decided that she would just have to outlast him. No matter how often he knocked her down, she kept getting up and would not give in. Eventually he tired and conceded defeat, becoming her friend and ally.

Youth and class were on Jeanne's side. Not all young men on the ship had fully developed the masculine traits that mature and fill out in older men. Furthermore, a valet to a Genevan or Parisian gentleman might naturally be expected to be a little more effete in his mannerisms than a rustic sailor. She was unlikely to be challenged to fisticuffs, but she carried two pistols in her belt even so. She was prepared to defend herself.

I imagine that when she presented herself to apply for the position as Commerson's servant in Rochefort, Jeanne appeared as a quiet and serious young man, already dedicated to, and knowledgeable about, botany and science. Altogether a suitable boy.

On her arrival in Rochefort, Jeanne would have smelt the salt and seaweed tang of the coast – her first taste of the sea. Rochefort was a beautiful city, which offered 'an infinity of things for an observer to see'. It was as white and clean and uniform as Paris was dark and twisted and crooked. There was no mistaking it for anything but a naval town. A partial star fort on a long loop of the Charente River, the naval shipbuilding complex was impenetrable from the sea, unassailable from land and protected from all weather. Parisian critics complained about the surrounding swamp, the mud and the unhealthy airs, but a little bit of water never bothered a sailor. Everything had a parade-ground neatness to it. The buildings were long and straight, with regular windows along their lengths. The roads turned sharp left or sharp right with military precision. Nothing stepped out of place.

After a century of neglect, modern Rochefort has been returned to its former glory, its historic buildings restored, renovated, cleaned and rebuilt, revealing their sharp pristine lines and crisp paintwork. Christèle and I drive through the busy streets towards the port in the centre of town. Christèle has not been here since seaside holidays as a child. She does not remember it being a very exciting place.

Our accommodation is moored at Quai aux Vivres – the dock where the *Étoile* would have loaded rations from the warehouses, which are now being renovated into apartments. We are staying on board a beautiful little Dutch sailing barge built in 1896. As I clamber down the companionway into the main cabin all neatly fitted out with carved wooden fittings and homely creature comforts, I feel as if I have stepped into a manifestation of my childhood dreams. And even though we never leave the dock, the gentle rock of the boat beneath my feet, the cheerful camaraderie of neighbouring boaties, the slap of rigging in the night and the intricacies of marine stoves, toilets and showers bring back childhood memories and make me long to cast off the mooring lines, hoist the sails and slide off down the Charente River towards the sea.

The *Étoile* was expected to be ready by December but the authorities at Rochefort did not consider her mission to be a priority. Bougainville would have to wait in South America for his consort ship. It would be February before the *Étoile* departed.

'The Rochefort Intendant, Mr de Ruis-Embito, has placed in the way of outfitting every difficulty and obstacle he could, and an intendant can do a great deal of harm when he puts his mind to it,' Bougainville noted in his journal.

The *Étoile* was not a new ship. Originally built as a trading vessel, *Paulmière*, in 1759, it was purchased by the navy in 1762 and renamed the *Étoile*. About two-thirds the size of the larger frigate, the ship was much roomier and required far fewer men than the *Boudeuse* to sail. There were only 120 aboard, giving everyone a bit more space – when it was not loaded to the gunwales with supplies and goods.

Such *fluyts* were modelled on a Dutch cargo vessel and known as 'corvettes of burden'. They were specialist trans-oceanic voyagers, with no adaptation to military service. The Dutch invested in trade, not war. The flutes were cheap to build, required fewer crew and could carry twice as much as other similar-sized ships. They played a significant role in Dutch supremacy over long-distance trade routes. Traditional flutes have flat bottoms and near vertical sides, as square as the limitations of steamed ribs and planks will allow. But the *Étoile*'s mid-section revealed a gracefully sloping hull beneath the waterline, designed to improve sailing performance as the ship heeled. The ship might have been beast of burden, but it still displayed the innovative flair of French maritime engineering.

Rochefort was a sophisticated production line for ship preparation. At the far end of the riverbank were the shipbuilding yards themselves, where bones of oak rose in various stages of keel, frame, ribs, planking and sheathing from within the stone crucibles stepped into the river bank. Six thousand trees, each 300–400 years old, per ship.

Vats of 'Norway tar' anointed acres of rope and filled the air with its smoky aroma. The forges glowed red with constant fury, producing ironwork. The flat horizon of the riverlands stretched around them, punctuated by bristling masts in all stages of dress. The great spars were lifted diagonally alongside the mast-raising pontoons, their standing rigging draping limp from the crosstrees. Once they were upright, tiny figures scurried spider-like along them, weaving their web of rigging – ratlines, halyards, clew lines, braces, reefs, lifts and tackles – until the ships were fully attired. Each quay along the river offered a different service for the new ships – like a specialty street of tailors where one might buy shoes, trousers, coats, scarves, hats and pipes at different shops. Here the ships progressed though masting, rigging, ropes, sails and stores until they could sail out to sea, in full regal dress, like a flotilla of dignified matrons across a ballroom floor.

It was the Royal Corderie that dominated the complex – then as it does now – a long low white building with elegant classical repetition. Its foundations 'float' on an oak raft on the unstable ground – an engineering masterpiece. The long narrow halls allowed the production of reams of hemp and linen rope. There are more lineal metres of rope on a ship than timber. The Corderie was as vital for naval success as the oak forests or the blacksmith's forges.

Some ropes were as thick as a man's arm and heavy, even when not laden with water. Jeanne would have recognised the equipment used to coil and twist them as a gargantuan version of a peasant woman's spinning wheel. Giant metal combs scraped and straightened the hemp fibres. The fibres were separated, then spun onto cotton reels the size of a loaf of bread. These reels of thread, some 30 or 40 of them, were mounted on a shelf of spindles and all spun together into a single cord. It was an impressively complex process. The straightening and smoothing work that Jeanne would have done with delicate precision between the thumb and finger when spinning, must here take six or seven fully grown men all their strength as they hauled on recalcitrant lines that bucked and twisted their objection to being harnessed into servitude.

Even modern ropes are wild and dangerous creatures, twisting against their inborn tensions and whipping with sudden viciousness. Ever since I can remember I have been warned about the perils of ropes entangling the unwary and dragging them overboard or crushing them in a deadly embrace. Never wrap a rope around your hand for better grip, my father warned me. Better to lose the load than lose your hand.

When we finally departed from Port Adelaide on the first stage of our travels around Australia, my grandfather travelled with us. We spent a rolling night sheltering behind Kangaroo Island, before continuing to the tiny safe harbour of Robe and then to Portland, across the Victorian border. We struggled to manoeuvre alongside the wharf in the busy harbour under the critical gaze of watching fishermen. Dad threw the boat into reverse to try a new approach and I watched in horror as a stray mooring line snaked up my grandfather's arm and pinned him helpless against a bollard. I could see the heavy nylon rope biting into his soft brown skin as I shouted in panic. Fortunately the boat had little momentum and he was quickly freed with no more than some nasty bruising. The incident certainly confirmed my grandmother's belief that boats were far too dangerous to live on.

I often think about ropes when I'm writing: the threading of multiple narrative strands through a single work with differing perspectives. How to spin such epic tales that stretch across voyages and lifespans, fraught with the challenges of maintaining a continuous unbroken ply of fibres without loose ends or frayed threads? I hate frayed ends. I have always derived great satisfaction from sealing the ends of nylon ropes over the stove and pressing the wax-like plastic into a rag until it cools as a smooth neat lump. Hemp ropes had to be traditionally whipped with tarred flax twine wound neatly around and around for several inches before being sealed off and securely fastened. Such binding was unbreakable. The fishermen in Port Lincoln often laughed about *Caliph*'s rigging 'held together with bits of string'. Their sneers seemed justified one day when, lifting a heavy load on board, a shackle in the rigging snapped. But the rope whipping was unbroken – it was a welded metal bracket that had snapped under the strain.

I wish I had learnt to splice ropes with a flashing marlinspike, seamlessly weaving and knitting the strands back against each other. Writers often use the metaphors of braiding and spinning to describe their work, but I am starting to feel that writing this book is more like ropemaking: a test of strength that threatens to overwhelm me in a tangle of serpentine twists. The customary skill of a practised hand may not be enough for this fibre, so much as the brute force of will, pitting the twists of each strand against one another to create a durable and seamless cable that folds in reluctant coils at your feet.

So many feminine crafts are amplified and made masculine on ships. The delicate work of the seamstress takes on gigantic proportions. On the banks of the Charente in Rochefort, acres of golden hemp were cut and layered, hemmed and stitched, edged and finished. My mother sewed *Caliph*'s brown cotton duck sails on an old Singer sewing machine in the hall of the local high school where she worked. Layer upon layer of fabric stitched together to make the wide sheets. Endless miles of stitching with giant needles pushed through stubborn cloth with a thickened leather sailmaker's palm. Brass eyelets were pounded into gleaming rows across the width to take the reeflines used to shorten the sails in high winds. Heels and heads were reinforced with six, eight or ten layers of cloth, every edge corded with rope to take the full weight of gale-force winds.

At last, finally, when the ship was equipped with all the necessities, it was time to load such supplies as met human needs. Novelist and botanist Bernardin de Saint-Pierre, heading for Mauritius a year or two later, provided a vivid account of the loading and departure of such a ship.

'I have given you a general sketch of the disposition of our ship,' he wrote, 'but to describe the disorder of it, is impossible.

There is no getting along for the casks of champagne, wine, trunks, chests and boxes everywhere about. Sailors swearing, cattle lowing, birds and poultry screaming upon the Poop; and, as it blows hard, we have the additional noise of the whistling of the Ropes, and the cracking of the timbers and rigging as the ship rolls about at anchor. Several other ships lay near us, and we are deafened by the hallowing of their officers to us, through their speaking-trumpets.'

The *Étoile* was no different. And into this chaos, stepped Jeanne.

It is a bit of a mystery what Jeanne looked like – either as a man or as a woman. She was small and strong, a little on the plump side, and apparently not particularly striking in any other way people thought worth mentioning. I wish there was an illustration of her standing on the deck, a quick portrait by one of the crew – like those done by Charles Alexandre Lesueur on Baudin's voyage or Louis Le Breton on Dumont d'Urville's. Many of the later voyages employed talented artists to document the voyage and its collections, but not this one. Commerson would have to do his own drawings at sea.

There are no known images of Jeanne drawn from life. A picture of her was created in 1816 by Giuseppe dall'Acqua for an Italian account of James Cook's voyages. In this full-length engraving, she carries a bundle of plant material limp over one arm, a motif for her occupation. Her plain face gives nothing away, discreetly looking down and away from the viewer. She does not look particularly short or plump, nor curvaceous or freckled. There is no sense of individuality in the picture. It's just a placeholder, a symbolic representation of a character created well after her death by an artist who never met her.

Dall'Acqua dresses her in loose sailor's clothing, characteristic of the times – shapeless and androgynous. A red woollen cap covers curly hair, a short navy blue peacoat opens to reveal a pyjama-like striped shirt and long wide-bottomed pants divided by a broad sash around her waist. She wears white stockings and black court shoes, the kind that I know were issued to the crew of Kerguelen's exploration voyage a few years later. The image is in the same style as many other dall'Acqua ethnographic engravings representing people in their 'native ethnic dress' for the travel books he illustrated. These are not portraits of individuals, they are 'types' symbolising a race or a culture. Or, in Jeanne's case, an anomaly.

I'm not an expert in French naval clothes, but I don't think there were regulation uniforms for French sailors before the Revolution. Red caps usually signify convict labour. I search through the series of hugely detailed paintings of French ports by Claude-Joseph Vernet painted in the mid-eighteenth century. For the most part, the workers and sailors in the port wear knee- or full-length pants, shirts and jackets in varying shades of white and blue.

And in any case, Jeanne was not a sailor and she was never part of the crew. She was a gentleman's valet or a manservant. Valets wore neatly tailored stockings and breeches, shirts, waistcoats and fitted jackets, in a simpler, less grand version of their masters, perhaps even their master's old clothes. Her origins might have been the lowest stratum of a society but, as a servant, she had attached herself to a higher level, and would almost certainly have dressed according to her new-found status, well above the sailors. The clothes make the man, as they say, or, as the French say, the habit makes the monk. I don't believe that Jeanne ever dressed like a sailor.

When Rose de Freycinet stole aboard her husband's ship as it left for a voyage around the world in 1817, she cut off her hair, dressed in men's clothing and boarded at night, grateful for the clouds. She was afraid of being recognised, despite her disguise.

'Before leaving the port we had to stop at the exit to give the password and somebody brought a light and I did not know where to hide myself,' she wrote. Someone asked her who she was and a friend quickly answered that she was his son. 'I was extremely agitated all night, thinking all the time that perhaps I had been recognised and that the Admiral-commandant, having been informed, had sent orders to put me ashore. The least noise frightened me and I was afraid until we were outside the harbour.'

Both Rose, and her husband, had much to risk by discovery. But once safely at sea, she had the security of her husband's protection and the privacy of the captain's cabin. When they were far enough away to avoid the navy's wrath, she opened her considerable trunks and returned, with obvious pleasure, to women's fashions.

Jeanne's path onto the ship was less troubled. Once she had convinced herself, and others, that she could pass for a young man, there was no reason for her to sneak aboard in the dead of night. She boarded as a legitimate member of the ship's company, and no doubt with much work to do getting all of Commerson's equipment and supplies successfully and suitably stowed on the increasingly crowded vessel.

Once Jeanne and Commerson were on board, they were allocated sleeping quarters. Commerson was given 'the most beautiful and most convenient of the vessel, without taking into account that of the captain himself'. This was not to say, however, that the accommodation was luxurious. The cabin had barely the dimensions of a household closet, with hardly headroom to stand, and was shared with other officers. It seems likely that this space would also have been shared with the officers' servants and Jeanne.

It must have been strange to embark on a ship of 120 strangers, none of whom you had met before, and none of whom you would be able to escape. Jeanne would be stuck with them for over a year, in a very small space, under difficult and testing conditions. It was not so much the sailors, the *matelots*, that Jeanne needed to be worried about. The crew of the *Étoile* were, like their officers, mostly rough-tongued, hardworking Bretons, who had their own distinctive and conservative perspective on the world. They had a well-established ecosystem in the focsle, from solid, unwavering helmsmen to the fearless, agile topmen. Sorted into watches, they followed the lead of their boatswains and leading seamen, accepted their orders and their punishments. The carpenters, caulkers and sailmakers protected their shipmates with wood, oakum and thread, the gunners with cannon and the fusiliers with rifles. Each tribe kept to itself and everyone knew their place.

But the stern quarters of the ship were a very different matter. The officers and passengers were a potent mix of pre-revolutionary France, with all the aspirational social tensions and conflicts between merchants and nobles, city and country, sailors and civilians, untutored and educated, tribe against tribe. Everyone had their status, their position in the hierarchy, and there was a constant battle to either maintain or improve it. Nobles thought they outranked civilians, gentlemen thought they knew better than experienced sailors. Surgeons bristled against naturalists who were, after all, just glorified medical graduates.

The surgeon on the *Étoile* was François Vivez. Vivez had been sailing since the age of seven, with his father who was senior surgeon on the *Formidable*. He knew far more about sailing ships, their crews and their ailments than a hopeless landlubber like Commerson. But Commerson was a highly regarded medical practitioner, had been appointed as the King's Naturalist and was paid more than *Étoile*'s Breton captain. I suspect this relegation did not please Vivez.

Others passengers and volunteers also occupied ambiguous places in the naval hierarchy. Some of them were important, depending on their standing or temperament or relationships. Pierre Duclos-Guyot was the younger son of the captain of the *Boudeuse* and sailed with his brother Alexandre to South America before transferring to the *Étoile* for the remainder of the voyage. Then there were the butchers and bakers, the musicians and the ship's boys – dozens of these, mostly stowaways and deserters. The captains of these ships, unlike ships of state, were obliged to feed and shelter all those on board until they could be put ashore.

And finally, there were the servants. Each of the officers had a servant, as did Commerson. Bougainville had four. Not many people write about the role of servants, on land or at sea. In the English navy, the captain's servants were not so much staff as sinecures allowing him to provide for old friends and family.

'A captain in the old time frequently put to sea with a little band of parasites about him,' wrote John Masefield in his depiction of sea life in Nelson's time. 'They stuck to him as jackals stick to the provident lion, following him from ship to ship, living on his bounty, and thriving on his recommendations.'

Ship's boys as young as thirteen often acted as officers' servants, although they were effectively training to be either sailors or officers, depending on their social background. Such servants played complicated roles as trainees as well as domestic staff, moving between the professional and domestic, crossing social boundaries between classes, as well as between the lower deck, the wardroom and the captain's cabin. Whether they were the sons of aristocrats or homeless stowaways, they were privy to many secrets of the ship. The role of a servant – amorphous and variable – was the perfect foil for someone in disguise like Jeanne.

The *Étoile*'s muster roll has never been found so we don't know how many servants were on the ship, or who Jeanne's peers were. They may have been subservient and polite, but they occupied the positions of privilege and knowledge attached to their masters, for better or worse. They were bound by naval regulation, but, like personal servants within households, also sat slightly outside the usual hierarchy.

As a servant, Jeanne's role would have been a mix of domestic and professional work. She would certainly have been required to carry out all of Commerson's domestic labour: washing and repairing his clothes, making his bed, fetching candles, keeping his cabin tidy, ensuring that he received his meals. These were generic tasks required of any servant, but they were also specific to Jeanne. Vivez described her washing clothes, and noted that she was good at sewing and had dexterous hands. Keeping Commerson fed and his goods tidy was no easy feat, given his preoccupation with natural history and neglect of his own wellbeing. Jeanne, unlike some of the other servants, was not training for a naval career, but in effect for that of a naturalist, or a naturalist's assistant. Her work would probably have been more onerous than that of most servants.

Just before the ship was about to leave, Commerson added a casual postscript on a letter to his brother-in-law.

'I must warn you in advance,' he wrote, 'not to believe any of the rumours that might spread after my departure. There is nothing so common in seaports as spreading false stories about the ships that have left.'

Perhaps he meant stories of ships sinking, being overrun by pirates or mutinous crews. Or perhaps there were other shipboard rumours that Commerson did not want his brother-in-law to hear.

7

ALL AT SEA

The Atlantic, March–April 1767

THERE IS NOTHING QUITE like the sensation of wind filling the sails: that sudden lift and sweep of the hull beneath you as if an inert floating object had unexpectedly come to life, risen from its slumber and set out on a journey of its own determination. All the random creaks and groans, the anxious tossing at anchor like an unhappily tethered mare, the loose slap of rope and canvas: all this uneasiness ceases as every fibre, every compass needle swings onto a single setting, and all the wild energy of wind and wave is harnessed into a swift and silent forward thrust.

It was a beautiful morning – fine and clear – when the *Étoile* finally sailed off anchor under the topsails and mizzen. In the light airs they were soon under full rig, raising as much canvas aloft as they could to drive the heavily laden vessel through the water.

The ship was chock-a-block with goods to trade – on the bridge, in the steerage, on the quarterdeck, in the large cabin – causing landlubbers like Jeanne, unaccustomed to the confines and athletics of marine life, to stumble, trip and curse as they accommodated themselves to their unstable home. But they soon found their feet, buzzing with excitement and praising themselves for their sturdy stomachs as the *Étoile* sailed north-east across the sheltered Bay of Biscay and believing themselves well attuned to the rigours of the ocean life ahead of them.

'I am already no longer a land-dweller,' Commerson wrote as they left the placid waters of the Charente River. 'I am writing this from the harbour off the island of Aix. My little test at sea was not painful: I think I shall enjoy a marine existence. I have not yet felt nauseous.'

His misplaced confidence is amusing. It is one thing to wear the right clothes, to talk the right talk, to step onto the deck of a boat, adjust yourself to its balance and confines and feel that that you are ready for a life at sea. When I was growing up, the stories of such new sailors were legendary. People set out on their sailing adventure of a lifetime, heading into open ocean, setting their autopilot, putting on their pyjamas and going to bed while their boat steadily sailed itself into the rocks. There is world of difference between looking the part and living the part. Commerson's seaworthiness was yet to be tested, and so too was Jeanne's manliness.

The open sea soon proved a different beast from the protected waters of the river mouth. The seasoned sailors watched as the ship began to roll under the first lift of swell. The gentle motion started like a rocking cradle, then increased in violence, with feints left and right as if intending to catch the unwary by surprise. As the ship pitched and swayed with all the fitful temper of a drunken sailor, the passengers surged to the gunwales, heaving in sympathy. Jeanne, like all the other greenhorns, was forced to empty her stomach of the customary tribute to the ocean gods, but still the deities demanded more. The unfortunate novitiates retched compulsively over the side with hollow cries, until their eyes blurred and their limbs collapsed from the convulsions.

'Eat, eat,' exhorted Commerson, between his own violent paroxysms. 'The convulsive efforts of an empty stomach will soon vomit blood and some pulmonary vessels could well break.'

But no food or drink tempted his own appetite, not even brandy. By the end of the second week, Jeanne had recovered and Commerson was the only passenger still ill, still vomiting blood. He began to doubt the wisdom of them having embarked on this voyage. Perhaps the ship was a trap. Someone had set the tasty morsel of bacon to entice them on board, he moaned, but once the sails were raised, the trap snapped shut – there was no way they could leave and there was nothing to do but chew on the bars.

Jeanne would have had no time for such regrets. She had to keep her master fed and clean up what he could not keep down: washing, mending, cleaning, feeding, tending. The wound on Commerson's leg had now reinflamed, oozing pus and ulcerating. It needed regular cleaning and dressing.

It wasn't until March that Jeanne would have been able to go on deck more often with Commerson. They watched a sleeping turtle drifting past, heavy as a barrel full of water and covered in barnacles and small shells. They saw porpoises, but few fish, although the men said the waters usually seethed with bream. But soon enough, a couple of sharks were hauled aboard, which needed to be dissected, analysed and described.

It was an old sea myth, Commerson insisted, that these sharks were always accompanied by pilot fish. What possible value could such an inconsequential fish have to a large and impressive predator? And yet here they were, small black and white consorts that 'liquidly glide on his ghastly flank', sometimes four or five to a shark. Commerson accepted the proof but remained mystified. The pilot fish might benefit from the shark's protection and morsels that fell from its mouth, but what did the shark gain?

There was no doubting the loyalty of the pilot fish. They hung close to their large companion, rarely darting far away. Even when a shark was hooked and hauled aboard, the pilot fish followed it up, only falling away as it swung in the air. They rushed back and forth, greatly disturbed by their loss, and trailed in the ship's wake for days as if hopeful of their companion's return.

Jeanne might have had a clearer insight into this unusual relationship than Commerson. Those who have always had others tend to their menial domestic tasks are often oblivious to the value of such labour – until it disappears. The sharks open their mouths in passive acquiescence and allow the pilots to clean between their teeth. Without their loyal cleaners, the sharks would soon be laden with parasites, drained of energy and vigour.

The benefit of such relationships is always mutual.

The story of the expedition, and Jeanne's role in it, was told by a variety of participants. There are different accounts written by different crew members for different purposes. Some are purely functional – carefully dated daily accounts of the weather, the ship's course and major events. Others are personal and anecdotal, written long after the events. Many of them were intended, ultimately, for publication. Exotic travel narratives were popular in the eighteenth century.

A journal was a useful activity to while away the hours of a long sea journey. Many of the journals were rewritten several times, so multiple versions exist, telling of varying events. Their writers collaborated and consulted with each other. Phrases recur and repeat in different journals until it is unclear who said what and when. On many French naval journeys, the commanding officer collected (or confiscated) all the journals on a ship as official records, using them as sources for a combined 'official account' published with considerable state support. Some wrote their journals as letters, to avoid collection for the naval archives. Naturalists also published their scientific accounts, sometimes on their own, but mostly as 'atlases' appended to the official narratives.

Bougainville's highly readable and popular account of the voyage was published soon after his return. It rapidly went into new editions, was abridged and translated, excerpted and quoted. More recently, Bougainville's personal daily journal has also been transcribed, translated and published in English. Other journals have been retrieved from the archives and published in part or in extracts, sometimes in English. From the voyage of the *Boudeuse* we also have the journal of Nassau-Siegen (in three different versions); the account of the ship's *écrivain* or clerk, Louis-Antoine Starot de Saint-Germain; and that of the volunteer Charles-Félix-Pierre Fesche – whose journal was possibly written in collaboration with Saint-Germain and also includes annotations by Commerson. On the *Étoile*, the young Pierre Duclos-Guyot kept a journal, often in concert with Commerson. Commerson wrote various manuscripts and journals as well as many letters, as did the surgeon Vivez (revised over time). The *Étoile*'s second-in-command Caro kept a fairly cursory sailor's journal – probably one of the few not pitched at publication, but sadly, the captain's journal by La Giraudais, which might provide a more objective account of some of the more contentious events on the journey, has never been found.

Jeanne warrants a page or two in the narratives of Bougainville and Vivez, is mentioned fleetingly in the journals of Saint-Germain and Commerson/Duclos-Guyot, disappears from successive versions of Nassau-Siegen's account, and is entirely absent from the rest. It's not a lot to go on. The record is both noisy and patchy. It's like trying to follow a conversation at a party where you only hear snatches of what's going on. It matters who is saying what and when, who saw things firsthand, who is repeating what others have said, and who might be just making things up for the sake of a good story. The journals say as much about their writers as they do about their subject. I can glimpse interactions, am never quite certain what is meant by the tone and the timbre, but I have to try to make sense of what there is anyway.

One thing is clear from all these journals and accounts: in those written at the time – the dated diary entries – no-one except Commerson mentions Jeanne at all until after Tahiti. It is only in the retrospective accounts, written or rewritten after their return to France, that anyone mentions any earlier suspicions about Jeanne's identity. Maybe they always knew, maybe they suspected earlier. But if they did, there is no evidence that anyone said so at the time, not even in their private journals.

It was the ship's doctor, Vivez, who later claimed that Jeanne's care for Commerson, including sleeping in his cabin, 'was not proper'. It's a slightly strange claim to make. Sleeping alone on a ship would be far more unusual than sleeping in company. The sailors slept side by side in the focsle, separated at best by a few regulation inches. The warrant officers shared cabins, the ship's boys dossed wherever they could find space. Not even the captain was guaranteed privacy, often sharing the main cabin with his servants, separated only by a canvas curtain. Commerson initially shared his sleeping quarters with the other officers. If he was ill and needed attention, his servant would certainly have slept nearby – this was, in fact, the point of the canvas servants' berths. Even on land, in the 1700s men habitually shared beds with other men, and women with other women, servants and masters included. Privacy and single rooms were scarce.

I am not convinced that Vivez is an entirely reliable narrator. His narratives are much reworked, and crafted for effect. There are multiple versions. The different versions of his published story, written shortly after his return and then after publication of Bougainville's account, reveal the way he changed his story, particularly the sections about Jeanne. Like Commerson, Vivez often writes in metaphors and vagaries, but his journals have neither the erratic spontaneity of Commerson's journals, nor the dry navigational reliability of some of the officers. The historian John Dunmore notes that his descriptions of Jeanne Barret are written in 'rather bantering and slightly salacious style'. His double-entendres about Jeanne and Commerson's 'mutual attachments' and 'quiet periods of enjoyment' rapidly become tiresome, and I'm not sure how much credence I should lend them. And yet I cannot ignore them. Vivez's journals are the main source of information about Jeanne. However much he gilds the lily, I still suspect that his accounts are based on some kernel of truth.

Many of the challenges for women at sea are obvious. For one thing, washing and toileting on board was very much a public activity. The ship's head is the ship's toilet. On sailing ships they were traditionally no more than a platform each side of the bowsprit with slotted panels to allow the waves to wash away the soilage, positioned behind the decorative head timbers that culminated in the figurehead. They offered no privacy and all the crew used them.

Other women who wrote about their life as men at sea in the eighteenth century, like Hannah Snell or Mary Lacy, did not mention any particular problems with toilets or washing. They must have managed somehow. Perhaps discretion all came down to careful timing – or a bucket.

Sometimes the lowest tech solutions are the best. Even internationally famous climate activists have to make do with a bucket on Atlantic crossings. Greta Thunberg made the trip on a zero-carbon yacht, with reporters excitedly describing its solar panels and hydro-turbines, to power the electrics on board. The old-fashioned technology that has driven much human expansion around the world – sail power – was all but forgotten. Old tech, like a bucket, doesn't often make headlines.

Not all of the challenges to being a woman at sea are as obvious as toilets. Some of them are peculiar and distinctive to the sea itself – unpredictable to anyone unfamiliar with its culture. As they sailed south-west, the weather slowly warmed and steadied until, in late March, the *Étoile* crossed the equator.

The Baptism of the Line was a rite of passage decreed by gods and kings older than any of those in France. No naval or royal decrees governed these proceedings, no captain dared deny the crew this ceremony, but instead duly relinquished the deck to Neptune or Father Equator for the duration of the proceedings. The social landscape was levelled – no nobleman was too aristocratic, no gentleman too important to escape it, no matter how absurd he might consider the whole escapade. All men stood equal before the sea. To shirk from the ceremony would not only create suspicion in the eyes of the highly superstitious crew, but also cast a shadow on the safety of the ship. It seemed there was no avoiding it.

As soon as Jeanne, along with every other initiate, received a letter demanding her participation, she would have known what was in store. The roll was called to ensure that everyone was present. Father Equator descended from the topmasts with all his acolytes. Initiates could expect to be stripped, dunked and abused in this ceremony. Less favoured crew, even junior officers, might be dumped naked into the water-filled longboat and covered in soot. The foredeck frequently degenerated into a free-for-all until everyone was battered and bruised, soaked and half-drowned.

Every boy on the ship stood naked, covered with a black mix of oil, soot or tar and covered with chicken feathers. The rest of the crew dressed in sheepskin, with devil's horns, tails and claws. Some of them walked on all fours. Others danced like bears. The boys swung in the rigging like monkeys. All of them were neighing, growling, meowing and barking madly to the accompaniment of twenty goat horns and every pot and pan from the ship's galley.

This noisy intimidating parade circled the waiting initiates until finally Father Equator, in his triple crown, took his throne and called his demons to silence.

'Do you promise to do unto others as is done unto you?'

'Do you swear never to make love to a sailor's wife?'

As the horned and decorated officials passed along the line, they dispensed their dubious honours. Soon they came to Jeanne. Was this to be her dénouement, her adventure cut short, exposed before she had even reached their first port of call? Perhaps, if Jeanne had still been an impoverished peasant girl, vulnerable to the fickle tempests of superstition and tradition. But she was not. She was a gentleman's valet now, with privilege and money on her side. And Commerson had already 'greased the fat pig', paid off the sailors and ensured a safe passage for them both through this potentially risky and humiliating ceremony.

'All we had to do was to place our offering in the bowl held by the other assistant,' he confessed, 'but we had agreed among ourselves that we would do this only after having been well and duly baptised, as we did not want to refuse the lavabo, and had prepared for this. We had merely come to the private arrangement that we would be spared the bawdy jokes reserved for the sailors, so that we were all well-watered, firstly by the officials of his ceremony, then by ourselves, time and again throwing bucketfuls of water at each other and sparing no-one.'

Commerson dismissed the superstitious event as depressing and barely even worth mentioning in his journal, for all he describes it in detail in a letter home, but Jeanne must have been profoundly grateful that they had been able to buy themselves out of the worst of it by means of a generous offering.

Commerson's writing often makes me laugh. His impatient enthusiasm, his furious irritation, his effortless superiority is laid bare in his writing. I can see why he must have annoyed some of his colleagues and captivated others. His circumlocutions and omissions drive me to distraction as often as his extravagant claims amuse me.

I order his journal from archives in the Muséum National d'Histoire Naturelle and it duly arrives, a large pale hard-bound volume so old and delicate that I worry that it will fall apart when I open it. If this were an Australian archive I probably would not even be allowed to look at it, certainly not without supervision, substantial cushions, gloves and having first been frisked for illicit pens. But eighteenth-century diaries are not so precious or rare in France. The supervising archivist delivers the manuscripts, returns to her desk and I am left to open the pages by myself.

The front page is grandly inscribed with its author and title in large ornate print. Each of the 380 pages in the book has had a border of four margins ruled in pencil along each side. Each page has been divided in half – four days per spread – enough for a journey of more than four years. The earliest pages are filled with Commerson's closely spaced handwriting, but the entries gradually reduce to a small paragraph for each day as they settle into the tedium, and sickness, of their first long sea crossing.

I am impressed that he managed to write anything at all during his two weeks of illness. I well recall writing enthusiastically in my own brand-new diary as when we first headed out to sea. My zeal declined as quickly as the swell picked up on passing the headland. I have never been a good diarist even when I was not seasick.

My mother often laughed about my fondness for 'making books' as a child. I would carefully decorate the front cover, print out the ornate title, inscribe my own name and write the blurb on the back. I was good at title pages, and the table of contents, numbering the pages and making a start on the introduction, but I almost never got further than that. The only one I finished was a book on 'Animals' for a school competition, although I remember the final pages being something of a trial. My mother was firm and encouraging – she was always one for finishing what you start. When I completed the book, she typed, printed and 'published' nine copies for my ninth birthday. Neither she nor I could ever have predicted how closely my adult career would follow my childhood interests. Perhaps the Jesuits were right when they claimed that if you gave them a child until the age of seven, they would show you the man. Parental guidance exerts a lasting influence, and every time I think about giving up on a book, even this one,

I remember my mother's rebuke and labour on determinedly until it is finished.

In the library of the Muséum National d'Histoire Naturelle, I look at the pile of manuscript boxes I have ordered from the archives. It is unexpectedly hot for early summer and the windows of the library open out into the famous gardens beyond. I would rather be outside than in this stuffy room, struggling with Commerson's handwriting, checking for passages and annotations that are not included in the transcribed printed versions. On 27 February they sighted land – the Canary Islands – and Commerson enthusiastically drew an outline of the coast. On 13 and 14 March there was a sudden flurry of activity when they caught the sharks and saw the pilot fish. Commerson marks the margin with a manicule: the classic pointing hand reminiscent of Monty Python sketches.

His writing provides a visual map of the ebb and flow of a long sea voyage – the long stretches of tedium punctuated by short periods of intense activity, and sometimes terror. A few days later he included a small pencil sketch of a tuna. His journal was starting to look like a biologist's. But his enthusiasm soon waned. By the end of March the entries reduced to brief headings, ceasing altogether on the first of April. They were still a month off the South American coast.

The remaining 250 pages of the journal are completely empty.

Historians rarely seem to mention the weather. I suppose weather doesn't usually have much influence in the archives, beyond the minor discomforts of inadequate heating or airconditioning. Writers too suffer from this pitfall. The weather is a vista outside my office window into which I prefer not to venture unless the conditions are perfect.

But weather is everything and inescapable for sailors. They watch the horizon, check the colour of dawn and dusk, trace the patterns of clouds across the sky, measure the rise and fall of the swell, and observe the behaviour of small creatures above and below the surface. They are anxious and always waiting for the weather, for the change, for the storm, which always comes one way or another. For sailors, the weather is wind. The sea moderates the extremes of temperature. Rain provides fresh water, but it is more important for dampening the wind – helpful in bad weather: devastating in the doldrums and monsoons where flat windless seas leave ships drenched and drifting under relentless rain for weeks on end.

It is the wind that dominates a sailing life. Wind fills the ship's sails, drives them forward, gives them motion and power. Too little and they drift without direction. The wind beats the ships back onto a lee shore, drives them sideways, rips sails from masts, snaps stays and spars and flings men from rigging. It is the wind that rouses the surface of the ocean into giant fists that pound and crush, that slices icy sheets of water off the surface and whips them stinging across skin and eyes. Nothing else much matters but wind – what direction it comes from and at what speed.

Land dwellers care more about temperature. I remember standing stoic on the back deck of our boat, before we left Port Lincoln, encased in oversized wet-weather gear, a garbage bag of dry clothes scrunched in one hand, my school bag in the other. We watched the white caps rip across the bay, the boats swinging wildly on their moorings, and listened to the wind screaming through rigging.

We waited for a lull, a pause, a space long enough for us to dash to shore, and not be swept out to sea, a moment where oars might row faster and stronger than wind.

'Now,' my father said, leaping into the dinghy. We tumbled

after him, dropping fast and sitting low, clinging to the side-board. Casting off, the oars pounded the water in short sharp stabs, giving the wind no purchase on their blades, maintaining our grip on the water beneath. Reddened faces slapped with cold salt spray, gasping breath. One hand for the ship, one hand for yourself. Eyes up, keep alert. There is no hiding from the weather at sea, no time for fear or anxiety. There is just surviving.

The wind rose and we flung ourselves behind a moored fishing boat, clinging in the shelter of its stern until the squall exhausted itself. We cast off and tried again, leapfrogging, boat to boat, until at last we reached the grateful shore.

In town, down the street, people buttoned their coats, pulled down their hats, leapt into their cars, went to work, carried on their business. They barely noticed the weather. Kids hurried through the school gates, shouting and laughing, excited by the wind whipping leaves from the trees, while we shook the salt and adrenaline from our veins. There was no danger here.

Recently I heard a fisherman on the news being interviewed about a cyclone in north Queensland.

'I've been at sea for 45 years,' he said. 'Every morning you get up knowing that Mother Nature will try and kill you.'

Bougainville's journal does not record the temperature, or the humidity, or the time of sunrise and sunset. Sun or snow make little difference. He only recorded the wind and waves. The breeze was fresh, light, good, slight, little, squally, variable, head, calm. The weather was fair, overcast, hazy, stormy, unfavourable. Occasionally, when they were in the tropics, he made a grudging mention of great heat or torrential rain.

I rather suspect the log books of our own voyage would be similar. Pragmatic rather than descriptive. My parents rarely recorded the things that were of most interest to me. I had hoped I might be able to refer to them, but they have been lost over the years, perhaps irretrievably damaged by heat and humidity. But perhaps it is better to rely on the imperfectly personal memories after all. Perhaps it is not necessarily what happened that is so important as how it affects us.

The *Étoile* was supposed to rendezvous with the *Boudeuse* at the Falkland Islands, tucked just north-west of the tip of South America, where Bougainville waited in vain through March and April 1767. The *Étoile* had intended to leave in December 1766, but did not get underway until February of the next year. They arrived off the coast near Rio de Janeiro, but took a further three weeks to sail just halfway – some 2000 kilometres – down the coast towards the Falklands. The *Étoile* was running short on fresh water and food.

The officers may not, in any case, have been in a hurry to meet up with the expedition commander – not until they had dealt with some of the private supplies they had loaded for their own profit. Such *pacquotilles* were absolutely forbidden on naval vessels, although in practice a blind eye might be turned to minor goods.

'To push it to the extent that the provisions are substituted, and food and water compromised for the unhappy crew, to deprive them of the space of the decks and oblige them to sleep here and there on badly stowed chests and bins in danger of being crushed by them, is an abuse,' declared Commerson. He clearly did not consider the private goods he had brought on behalf of a Rochefort surgeon, to be sold at a profit of 628 livres, to be the same thing.

The *Étoile* finally sailed into Montevideo for repairs and resupplies on 30 April 1767.

If there had been rumours during the Atlantic crossing, if they had been investigated and substantiated, then Jeanne could have been sent home from South America. Several women were returning home from the colonies. Jeanne could easily have been affixed to one of these families as a servant.

But she was not.

Vivez claimed that at some point before their arrival at Montevideo, Jeanne had been forced to sleep with the other servants after complaints had been made about her sharing Commerson's cabin. The servants, apparently, believed she was a woman and wanted the matter resolved. According to Vivez, Jeanne told them she was a eunuch and, after a time, for lack of further evidence, interest in the matter died down.

No-one else mentions this particular incident, but Saint-Germain, Bougainville's naval clerk on the *Boudeuse*, later claimed to have had his doubts about Jeanne.

'It had been long suspected that M. de Commerson, a botanist doctor aboard the *Étoile*, had a girl for a servant, whom he had embarked at Rochefort,' Saint-Germain later recalled in his account of the journey. 'At Montevideo, there had been much discussion; which various sailors had wished to visit. But the captain, who, I believe, was not interested in secrecy, caused the most severe prohibitions to be made in this respect.'

Saint-Germain assumed that Commerson did not know Jeanne was a woman when he employed her in Rochefort. 'But he knew it in Montevideo,' Saint-Germain notes. 'They even had fairly conclusive proof of that. Why did they not send her back from Montevideo with the Falkland colonists?'

Perhaps neither Bougainville nor La Giraudais would have wanted to upset their prized naturalist by sending his servant home. Commerson received lucrative offers in South America to travel across the continent and rejoin the ships on the other side. In a land of such obvious biological wealth after a long and nauseating sea voyage, it must have been very tempting.

Saint-Germain does not discuss what the 'fairly conclusive proof' in Montevideo was. But Vivez does say that Jeanne fell ill during their stay at Buenos Aires, on the opposite side of the Rio de la Plata from Montevideo. Perhaps with his privileged medical knowledge, Vivez had some evidence.

'Scandalous gossip claims that she suffered at Buenos Aires from an acute illness brought about by the care she gave her master to relieve him from the weaknesses he might have had during the nights when she watched over him.'

An acute illness, particularly one caused by spending the nights with Commerson, sounds suspiciously like Vivez is hinting at a sexually transmitted disease. Such diseases were rife on eighteenth-century ships and, as the ship's doctor, Vivez was responsible for treating them, just as he was in charge of treating the wound on Commerson's leg. But Vivez reports Jeanne's illness only as 'scandalous gossip', not as a known fact, which makes me think she did not seek treatment from him. Other woman have avoided medical treatment so as not to reveal their gender. Hannah Snell said she removed a bullet from her own thigh rather than reveal her identity to the ship's surgeon. I would not be surprised if Jeanne avoided Vivez's ministrations. Commerson could have provided her with most medical treatment. Vivez's claim of a sexually transmitted illness sounds like just another opportunity to make sexual allusions. Accounts of women at sea are so often coloured by implausible accusations of sexual promiscuity that I am immediately sceptical. There is no other evidence that Jeanne was sick in South America, and in fact there is ample evidence that she was healthy enough to be extremely busy indeed.

I am curious about La Giraudais's 'severe prohibitions' though. Perhaps he decided to turn a blind eye to the rumours. It would not be the only time that a captain had chosen not to reveal the identity of a woman on their ship. Marie Louise Victoire Girardin, who sailed on d'Entrecasteaux's 1791–94 expedition as a male steward, was introduced to the captain Jean-Michel Huon de Kermadec by his sister. It seems likely that both Kermadec and d'Entrecasteaux knew her true identity.

Perhaps La Giraudais simply did not believe that the rumours could be true. Jeanne did not behave like a woman, and what woman would choose to sail aboard a ship anyway? One thing everyone agreed upon was that Jeanne was well behaved and caused no-one any trouble. And once they reached South America, her true value as a collector would be revealed.

8

GATHERING NATURE'S RICHES

South America, May–December 1767

It is strange to set foot on the unforgiving solidity of land after so long at sea. Some lightness of being is lost, like a bird that can no longer fly or a ballerina who can no longer dance. The swoop and sway of the sea, the skittering steps, the arboreal momentum vanishes, replaced with the leaden-footed regularity of one step after the other. Gravity changes and we are weighed down, reshackled to the earth, no longer free-floating and pelagic.

As the *Étoile* approached, the Spanish town of Montevideo appeared at the head of a wide flat bay: an isolated fortress rising in man-made mimicry of the mountain on the opposite point. The great 'river of silver', the Rio de la Plata – so wide it might have been another ocean – stretched to the south, fed by the twin waters of the Paraná and Uruguay rivers. Somewhere in the distance, in the heart of the estuary, lay Buenos Aires. But for now, the only clue to the rivers' existence was the drifting line of brown muddy turbulence that circled and swept towards them like tendrils from the earth seeking to extract them from the sea.

Montevideo was an impregnable star fort built on a peninsula jutting out into the sea, as if afraid that the forests beyond might encroach on the wide streets that ran in a rigid grid through the city. These South American settlements had long had reason to fear the inland. The local residents had fiercely opposed incursions on their lands. But the forces of European colonisation were unrelenting and unmoved.

As they stepped off the *Étoile* and walked through the streets of Montevideo, the visitors would have come under the shadow of the church that dominated the centre of town, its domed spires and flying buttresses brandished before the squat grey Cabildo on the opposite side of the square. The twin powers of church and state were surrounded by the strict hierarchical layout of segregated social and ethnic classes embedded in the Laws of the Indies that governed Spanish town planning in the New World. They had landed back on earth, to the normality and stability of an almost familiar terrestrial existence.

No doubt the goods that had been brought on the *Étoile* were promptly traded and whatever supplies they needed were rapidly obtained, and at a remarkably good price. This was a process Jeanne understood well – trade and markets, supply and demand, the exchange of coin for food, labour and goods. She would have found it incredibly cheap living in Montevideo as she visited the market for supplies. An ox cost only twenty pence and a horse just ten.

The country was an enticing temptation for anyone who had tired of a life at sea, as Bougainville observed.

'The first reflection which strikes [the sailor], on setting his feet on shore, is that they live there almost without working,' he wrote. 'Indeed, how is it possible to resist the comparison of spending one's days in idleness and tranquillity, in a happy climate, or of languishing under the weight of a constantly laborious life, and of accelerating the misfortunes of an indigent old age, by the toils of the sea?'

For the naturalists, for Jeanne in particular, the stop was anything but an opportunity for idleness. The forests beyond the city were a naturalist's treasure trove. South America was 'the most beautiful, the most temperate, the most fertile but also the most uncultivated part of the Universe.'

Commerson was faced with the 'the same perplexity of abundance as Midas, who saw gold formed wherever he laid his hand. Nothing less than the eyes of Argus and the bands of Briareus could see or gather the profusion of rare and hitherto unknown materials, that this country affords for extending the domain of Natural History.' But Commerson did not have Argus with 100 eyes or Briareus with 100 arms at his disposal. He had only Jeanne, who set out to prove her worth on the voyage by working as hard as three men.

I imagine her dressed like one of the naturalists on d'Entrecasteaux's expedition 25 years later, in a voluminous coat of many pockets, protective leather gaiters and a broad-brimmed hat. I've worn similar clothes myself for field work – most biologists do. Knives, guns, bags, pins, hammers, forceps and wads of linen were stock in trade for a collector and naturalist.

'During our stay at the Rio de la Plata,' Vivez wrote, 'she went to collect plants on the plain, up the mountains two or three leagues away carrying a musket, a game-bag, food supplies and paper for the plants, always 8 or 10 large sheets.'

Jeanne's equipment provides clues to her tasks. If she carried firearms she may also have been required to use them to obtain specimens. Perhaps she used the musket, loaded with shot, to collect the tyrant flycatchers – one black with prominent white spectacles and the 'austral negrito' described in Commerson's notes. Most gentlemen naturalists – like Banks and Darwin – considered themselves fine enough marksmen, but still were happy to have others shoot for them as well. Darwin tasked his manservant Joseph Parslow with boiling and cleaning the bones of a pigeon so that he could examine the skeleton. Even as a student he 'employed a labourer to scrape during the winter, moss off the old trees . . . and rubbish at the bottom of the barges in which reeds are brought from the fens' to obtain rare beetle species.

Darwin did not have a servant on the *Beagle* initially. It was only in South America that he realised how much help he would need. The captain arranged for Syms Covington, a fiddler and cabin boy, to work for Darwin. Covington had already hunted for rhea eggs and excavated giant megatherium fossils with the naturalist. Darwin's main intention was to train him to 'shoot and skin birds', but Covington was more than just a dogsbody – he kept a journal and his own collection. In fact, his bird collection helped establish the relationship between Darwin's famous finches and the individual islands of the Galapagos from which they originated. Unlike Darwin, Covington and other collectors on board were punctilious in labelling their specimens with exact locations, information Darwin would later use to analyse his own collections. Record keeping was the eternal bane of collectors like Darwin and Commerson – no matter their genius. At home the task of making sense of their collections proved a greater hurdle than the voyage.

Not everyone on board valued the collections of the naturalists. The fruits of Jeanne's labours may have been no more appreciated than those of Darwin and Covington at the time.

'Notwithstanding our smiles at the cargoes of apparent rubbish which he frequently brought on board,' the captain of the *Beagle* wryly noted, 'he and his servant used their pick axes in earnest and brought away what has since proved to be most interesting and valuable remains of extinct animals.'

Jeanne and Commerson had soon collected 20–25 boxes of specimens, notably stony and soft corals. Somehow I didn't expect them to have collected so much coral. I just don't associate the Atlantic with coral diversity. Although there are certainly shallow warm-water reefs in the Caribbean and cold, deep-water reefs even in the Arctic, it is really the Pacific, particularly the 'coral triangle' of South-East Asia, that is the epicentre for the greatest diversity of modern sun-loving, reef-building corals.

Cold-water corals were literally out of my depth as a beach-combing collector on the south coast of Australia. But when we stopped for a few months in Port Fairy, my dad worked on one of the crayfishing boats and, along with an excess of lobster that could not be sold, he would bring home bycatch brought up in the pots from the bottom of Bass Strait. Some of my finest shells came from this source – my largest cone, an asymmetrically weighted auger, the beautifully diverse southern volutes with vibrant mantles matching their striped shells, and a range of tiger cowries with their patterns mapping the depth and location at which they were found. Among them, too, were the delicate fans of gorgonian soft corals in orange and red, often with brittle stars still attached. They dried rapidly in the sun, retaining their vibrant colour as well as a slightly fetid odour, but it proved impossible for me to keep them safely. They were incredibly fragile, and invariably crushed, cracked or shattered over the course of our travels.

I can find no evidence that Jeanne collected shells in South America, even though Vivez described her collecting them several times in his narrative. Nor is Commerson associated with any shells that I have heard of, only plants, some fish, a dolphin, and the occasional bird or lizard.

'There were no shells from the Bougainville expedition,' the malacology professor at the Muséum Nationale d'Histoire Naturelle tells me firmly. 'Commerson was a botanist.'

I'm sure he's right about the museum's collections. But Commerson is known as a botanist because his plant collection survived. In truth, he was a naturalist. If he'd published his work on Mediterranean fish, he'd be better known as an ichthyologist. Or an ornithologist from his bird work, or an entomologist, or even a malacologist. Our vision of history is skewed by the evidence that has survived.

I'm sure they must have collected shells too. Jeanne was frequently described as collecting them. Shells from Bougainville's earlier voyages to the Falklands were among the earliest to appear in European collections, and many of the shells from South America were still new to science at the time Jeanne was collecting there. So where, after all this time, have they gone? Shells are surely one of the easiest things in the world to collect, the easiest to find, to clean, to store and transport. They are robust and durable, as their abundance in the fossil record attests. They were eminently commercial too, in the eighteenth century. They literally lie on your doorstep as you come ashore from a boat, as if waiting for you to fall over them. You'd have to go out of your way to avoid collecting them.

It was an opportunity ripe for a visiting naturalist to exploit. They would have collected the various species of river mussels growing in the shallow waters or myriad other bivalves: ark, wedge, saltwater or Venus clams, the bittersweets or dog cockles, oysters, glass scallops or tellens. If one had come to hand, they could not have refused a giant helmet shell or the goliath strombs so common to the north.

But I don't know what shells they collected, because their specimens never returned to France. The first set of cases Commerson sent back from South America with the Falkland Islands colonists never arrived at the museum. They were either lost or misappropriated. I only know they must have collected a white sea catfish from Rio de la Plata, because Bernard Germain de Lacépède later described this large and striking fish, mostly probably from Commerson's drawings and notes. After this loss, Commerson swore he would not let his collections leave his sight again.

It is ironic really that collections as robust as shells did not survive and yet the most fragile and delicate of pressed plants did. Almost 3000 plant specimens collected by Commerson and Jeanne on this voyage have been preserved and documented online. Of these fully 40 per cent came from South America. More than half of these, some 600 specimens, come from the Montevideo and Buenos Aires region. Jeanne was indeed busy around Rio de la Plata.

I can't imagine how difficult it must have been to keep pressed plants safe on a leaky timber ship. Damp, salt, insects and rats are the enemy of plant collections. Modern herbaria are kept in carefully humidity- and temperature-regulated buildings, with an exacting program of complete insect annihilation. The chronically damp, insect-ridden eighteenth-century sailing ships offered the worst conditions for preserving plants. The plant presses took up considerable space, needed to be kept in prime warm, dry locations (often in the already crowded officers' wardroom), and paper was at a premium and subject to chronic pilfering.

And yet, once dried, pressed and bound in loose volumes, perhaps wrapped in waterproof cloth and sealed in cases, the herbaria contained a vast amount of new knowledge that took up relatively little space. And with adequate protection they proved remarkably resilient. Specimens collected 200 or 300 years ago retain all the features necessary to identify and describe them. Their shape, structure and reproductive features are all noted on a single sheet of paper, annotated with date and place of collection, the name of the collector, the describer, and any particular details of the site or circumstances, depending on their owner's diligence. Other hands have added further details over time, revised names – new taxonomies, new homes – as specimens are transferred, swapped and exchanged around the world. Ultimately, each sheet represents a single specimen, encased within a cardboard envelope or folder, individually numbered and catalogued, and filed in a drawer with other representatives of the same species. The collector is all but irrelevant here. It is the plant, not the person, that matters.

And sometimes, in a particular institution, one of these paper sheets will be encased in an envelope of a different colour – usually orange or red. These are type specimens: the first specimen of a species to have been described, drawn and published by a scientist. Type specimens are literally the archetypes for their species, against which all other variants are assessed and compared.

A great many of the plants from Montevideo were type specimens. This collection was one of the first systematic surveys of South American plants, and among their number are both rarities and household names. Not all of the plants are dramatically exciting – the variations in the vast swathes of grasses and sedges might only excite a botanical specialist – but the roll call of specimens includes many species which belong to genera familiar even to the most casual gardener: *Abutilon*, *Amaryllis*, *Anemone*, *Aster*, *Buddleja*, *Cineraria*, *Convolvulus*, *Juncus*, *Lupinus*, *Oxalis*, *Petunia*, *Ranunculus*, *Rudbeckia*, *Senecio* and *Verbena*. Even dried and pressed on ancient papers for more than two centuries, their bright flowerheads remain bold and conspicuous, even if their colour has faded. Many of these are the ancestors of the flowers that adorn our gardens today.

From the testimony of others on the ship, and from Commerson's own admission that his leg was still causing him trouble, there is little doubt that the bulk of the South American collections were made by Jeanne. Botanists often recruited amateurs to collect specimens, and many of them conscientiously recorded the names of their collectors on their specimens, and this contribution is echoed in the databases and filing systems of modern taxonomy. But many did not, particularly when the collectors were Indigenous, or servants, or women.

Jeanne Barret is not listed as a collector on a single one of the thousands of specimens she obtained.

While Jeanne was busy collecting specimens for Commerson, Bougainville caught wind that the *Étoile* had arrived in Montevideo. He sent word from the Falklands, via some Spanish ships, that the *Étoile* should rendezvous with the *Boudeuse* in Rio de Janeiro. In due course, the two ships lay side by side in the harbour, and Jeanne and Commerson finally saw their famed expedition commander for the first time.

Bougainville was impressed by the scientists on the *Étoile* and invited them to transfer to the *Boudeuse*. The quietly capable astronomer, Pierre-Antoine Véron, accepted, but Commerson had reservations. He got on well with the officers aboard the *Boudeuse*, particularly Bougainville and Nassau-Siegen, but the *Boudeuse* was a frigate – fast and light. It was chronically short of space, so much so that the *Boudeuse* relied on the *Étoile* to carry enough food to support them on their long journey.

If Commerson transferred to the commander's ship, where he liked the company better, he would have had to leave his collections, his equipment and his books on the *Étoile*.

'I was obliged to ask, at our junction in Brazil, to stay on the *Étoile* which offered me the conveniences and ease that I would not have had on the *Boudeuse* where they could only receive me without my many accessories, which I could not accept without condemning myself to uselessness,' explained Commerson. 'Should I, for mere social amenities, give up the more convenient location that the captain had been obliged to let me have by orders from above, and to separate myself from my instruments, my books, my boxes?'

They decided that they would stay on the *Étoile*, in the company of officers whom Commerson bitterly dismissed as having 'never been more than cod fisherman and pirates of the state'.

I wonder if Jeanne had a more respectful relationship with the 'cod fisherman and pirates' than Commerson. I doubt that she would have disparaged La Giraudais and his fellow Breton officers for being mere mercantile 'trinket traders'. These men were generous and pragmatic with their awkward passengers. It was probably in Rio that La Giraudais gave up his spacious captain's cabin for Commerson, Jeanne and their collections. This cabin, more than anything else, may have been a major factor in them remaining on the *Étoile*. Without question it would have benefited Jeanne.

It is unlikely that Bougainville would have forced his captain to give up his cabin for the naturalist. If La Giraudais suspected that Jeanne was a woman, he would have wanted to separate her from the other officers. And the officers may have wanted to be separated from the annoying naturalist. It would also have been obvious, as soon as they touched land, that Commerson would rapidly fill any space he had with a wealth of natural history objects which, to be honest, few others would be keen to live with.

On every French scientific voyage, the naturalists struggled for the space they needed to do their work – not only warm dry areas to press their plants, but also space to dissect, measure and describe fish before they were sent to the galley for dinner, room to draw and paint those specimens whose soft tissues could not be preserved, areas free from vermin to strip and clean the flesh from bones, boards on which to pin and tan skins progressively stripped of fat and sinew. And even then, dried, pickled, skinned and stripped, they needed space to store the precious jars, boxes, crates and satchels ready to send home. No matter how well prepared and preserved, the aroma of decomposition does not readily leave these specimens. The smell was unlikely to bother someone like Jeanne, raised on a farm in close proximity to livestock, but few naval officers were keen to share their quarters with such delicacies and many could not see any value in them at all. Very often it was left to the commanding officers to intervene and ensure that the naturalists could continue their work.

Putting the naturalist in the captain's cabin solved a multitude of problems, and clearly the down-to-earth La Giraudais was not concerned with the social niceties of where he slept. In addition to solving the problem of Commerson's collections, it also had the benefit of providing Jeanne with a more secure living arrangement, not to mention access to a private toilet. The captain's 'privy' was in a tiny enclave at the end of the long line of windows across the high, overhanging stern. It was, in effect, a 'long drop' toilet, leaving any ordure in the ship's wake – and surely with one of the best views on the ship.

From here Jeanne's life on board must have become a little easier.

~

The Portuguese settlement of Rio de Janeiro was as beautiful as the rest of South America. And it was not just naturalists who thought so. Some 50 years later, Rose de Freycinet was enchanted by its unspoilt splendour, with cascading waterfalls that put the famed fountains of the Château de Saint-Cloud to shame.

'The immense forests,' she wrote to her mother, 'are still virgin and penetrated only by one narrow path barely wide enough for a loaded mule; they are exactly like the forest that our dear Chateaubriand described in *Atala*. The foliage is of a thousand different shades of green, lit by the brilliant blossoms that load the trees; a hundred sorts of creeper, each one more magnificent than the other and with beautiful flowers of elegant shape, bind the trees together and make it difficult to penetrate them; it would be delightful to push one's way right in and listen from there to the delicious music made by the thousands of birds, all remarkable for their beautiful plumage. It gave me intense pleasure to find myself surrounded by so much that was new and strange and I would gladly have stayed there alone if I had not noticed that everyone was moving on.'

But the forests Rose so admired were out of bounds for Jeanne. The Portuguese were so suspicious of the French intentions they restricted any scientific activities. Excursions outside the city limits were not permitted during their one-month stay. Véron the astronomer reported to Bougainville that the Viceroy would not provide facilities for making astronomical observations.

'M. de Commerson, the famous naturalist . . . is encountering similar obstacles,' noted Bougainville.

'You know in my fervour to see this,' Commerson wrote to his brother-in-law, 'in the midst of all the hostilities and in spite of a terrible leg injury which had returned at sea, I dared to go down with my servant twenty times in a little pirogue, under the direction of two negroes, and to traverse one after another the different coasts and islets of the bay.'

This 'servant' must have been Jeanne, one of the only times that Commerson explicitly acknowledges her presence. With his injured leg, Commerson would have directed most of the collecting from the boat, rather than scrambling ashore himself. It was Bougainville who had curtailed Commerson's activities. Vivez the surgeon had informed the commander of the risk of gangrene, that Commerson might lose his leg.

'M. de Bougainville . . . decided to improve matters by obligingly bringing me to a halt [*mettant aux arrêts*] until I was completely cured,' explained Commerson. His use of the word '*arrêts*' is sometimes interpreted as an arrest but it seems more likely to mean 'stop' or 'halt' in this context – more a means of protecting Commerson from himself, rather than a punishment. I don't think it means (as has sometimes been suggested) that Bougainville knew about Jeanne's identity and was punishing Commerson for his transgression. If Bougainville had known about Jeanne's identity in Rio de Janeiro he would have been obliged to send her home. And relations between Commerson and his commander remained mutually amicable.

Among the collections Jeanne made as they scoured this coast was a spectacular shrub that scrambled through the forest, adorned with vicious spines alongside the variable pink, red and purple bracts that surrounded its inconspicuous white flowers. Commerson duly named the plant *Bougainvillea* (now *Bougainvillea spectabilis*): a gracious and carefully considered gesture naming one of the most flamboyant and impressive plants they found in honour of their commander.

But Commerson's ill health and Bougainville's restrictions certainly took their toll on the collecting. Only 13 per cent of the surviving specimens collected from South America are from Rio de Janeiro, suggesting that Jeanne's days were nowhere near as productive as they were in Montevideo and Buenos Aires. Temporarily deprived of new specimens, Commerson turned his attention to Rio de Janeiro's other attractions.

'In the middle of winter, oranges, pineapples and other delicious fruits succeed each other in a rich variety; the trees are never robbed of their verdure, and thousands of slaves have nothing to do but gather the treasures, which uncultivated Nature lavishes in inexpressible profusion, furnishing man with the most delicious subsistence, without labour or industry. The mines of the country, which are numerous and rich, contain nothing but gold and precious stones, and a bay (twelve or thirteen leagues in circumference) formed by an arm of the sea, which abounds with excellent fish, offers a station capable of containing a hundred thousand ships secured from storms on all sides.'

I am a little surprised by the mention of the mines – and the slaves. The trade in Brazilian diamonds, after the decline of supply from India, was vast, and meant to be highly regulated by the Portuguese. But the port of Rio de Janeiro leaked illegal diamonds at its seams. Commerson, for all his indignation at the officers' illicit trading, was far from being above a bit of contraband himself.

'I am bringing you a sample of the diamonds from Brazil,' he wrote to his brother-in-law. 'Archambeau will have a little negro if he is wise . . . For myself, I hope I will not come back to you with empty pockets.'

Contraband diamonds and slave boys – that was not the kind of collecting I had been expecting.

When Véron the astronomer transferred to the *Boudeuse*, his place on the *Étoile* was taken by Pierre Duclos-Guyot, the younger son of the *Boudeuse*'s captain. Like Commerson, Duclos-Guyot kept a journal, but his was not a fancy dedicated hard-bound volume. Instead, he wrote on small sheets of paper, haphazardly pasted together at a much later date into a binding in which it does not really fit. The title page is a printed doctor's inventory with columns for maladies and medications. Duclos-Guyot's name does not appear anywhere – this was an aide-mémoire, not a manuscript for public consumption. He did not waste great sheets of paper on empty days but simply devoted a line to the weather with the dutiful regularity of a sailor or a farmer whose survival depends upon careful observance of the vagaries of the world around them. And from July 1776 onwards, his entries were annotated by Commerson, who clearly preferred to add marginalia or postscripts – sometimes on additional pages – to the careful regular documentation of his colleague. This diary provides another important source of information about what was, and wasn't, said about Jeanne during the voyage.

Sooner than they expected, Bougainville ordered the ships to depart from Rio de Janeiro, and on 15 July they sailed back to Montevideo to complete their preparations. But ill fate still dogged them. One windy night, a Spanish ship dragged anchor and drove into the *Étoile* with such force that it snapped off her bowsprit and carried away the knee and rails of her head. Somehow, in the darkness and wind, the crew managed to disentangle the two ships, but the *Étoile* was badly damaged. The ship had to be unloaded and moved up river to Ensenada de Baragán, where timber for repairs could be found.

The risk of dragging anchor is an ever-present anxiety with boats. Not so much when the weather is calm and the harbour sheltered, but when the wind picks up, causing boats to pitch and shy on their anchor chains like flighty horses against a halter. I remember being alone with the cat, doing schoolwork once, when I noticed a smooth sliding sensation, quite distinct from the regular tug and pull of the anchor. I ran to the deck to see the shore receeding faster than my dad could row towards us, as a bullet of gully wind blasted *Caliph* out to sea. There was no way he could catch up and if I didn't act we would quickly be in deep water. I had no choice but to let out the rest of the anchor chain and hope that the weight would slow or stop our progress. All 100 fathoms of chain finally halted the boat's flight, which we then had to painfully winch back in once my father caught up. For the early expedition ships, an added risk was losing the anchors entirely which could not be readily replaced on a long voyage.

Everyone on the *Étoile* would have been anxious to get underway, even though bad weather and unfavourable winds delayed their departure. But Jeanne kept up her work as they waited at Ensenada de Baragán for the *Étoile* to be repaired. Commerson recorded they saw a pair of little reddish brown birds that built mud nests and sang together in a sweet little trilling duet, although which one of the pair was male and which was female was hard to tell, they looked so similar.

Finally the weather improved enough for the ships to leave Montevideo on 14 November 1767 – a full year after the *Boudeuse* had originally left Nantes. But the weather wasn't done with them yet. No sooner had they set offshore than they were struck by a gale. The weather was deplorable and the heavy seas set the ships to unbearable rolling. All but two of the cattle on board were lost, *Étoile*'s topsail yard and chain plates were damaged, and they were still taking on water. It must have been an immense relief when the seas finally calmed enough for them to approach the tip of the continent and enter the relative protection of the Strait of Magellan.

For all the ferocity of the elements, the waters and skies here were filled with wildlife. Albatrosses soared above the ships, watching their progress with dark unmoving eyes. The crew murmured disconsolately at the sight of storm-petrels, whose presence never heralded calm waters ahead. And seas seethed with seals, penguins and huge numbers of whales, their skin engraved with white encrustations.

Ships are not the only floating armadas in mortal danger from such storms.

On the extremities of the southern continents – South America, Australia and Africa – it was not uncommon to find wreckage swept ashore in the aftermath of bad weather: timbers and buoys from boats, worn white and smooth from years at sea. In my childhood, even the trash was treasure: old ropes, mismatched thongs, frosted bottles. The frigid circumpolar currents locked around Antarctica protected the Southern Ocean from the excesses of our species, although no ocean on earth is safe from increasing pollution.

After a storm is the best time to beachcomb. The sky is still dark with rain, the sea grey and heaving from the storm. The wavelets hiss and crash against the sand in mock fury, as if remembering the violence inflicted through the night. The beach has changed shape, patches of sand gouged away revealing a pebbly underlayer. Kelp and seagrass pile up like some decomposing leviathan, where tiny fleas swarm and pounce.

The winter beach is almost monochromatically subdued compared to its blinding summer brilliance. The only colour rolls along the high-tide line, white spume flecked with purple and delft-blue, strangely brilliant and harmonious in violet hues. A tangled mass of creatures hard and soft, divergent in form, encompassing two of the oldest multicellular phyla of life,

Cnidaria and Mollusca, the wreckage giving little clue to events of their mid-ocean lives.

The transparent bladders of bluebottles puff and twist in the wind, anchored now by still vindictive tentacles. It is impossible to imagine from the tangled carnage how beautiful these chimeric creatures were in life, floating on a crested pillow, drifting tentacles 50 metres long, each studded with paralysing nematocysts that retain their fatality even in death. They are the original longline anglers of the oceans, trawling deadly and diaphanous hazards into the depths. I've been told that sea turtles, and sometimes marlin, eat jellyfish. But even sea turtles are not fond of the *Physalia physalis*. The tentacles sting them too – leaving them with swollen faces and eyes. There are more innocuous meals in the sea.

The wave of purple contains other creatures beside the bluebottles, strange transparent tunicates with their see-through barrel-shaped bodies exposing internal organs on permanent x-ray. By-the-wind sailors, *Velella velella*, hold their concentric crystalline sails aloft, set to tack either into or against the wind. Their deflated bodies are easily confused with bluebottles until they dry and degrade, leaving only their durable sails, like little triangles of plastic, to confuse beachgoers. Just rarely, there might be an exquisite blue dragon, *Glaucus atlanticus*, all flowery protrusions and brilliant colour and a powerful sting.

Piles of tiny purple snails stack up around the jellyfish, their shells so delicate they crush between your fingers. They are too fragile to survive contact with the shore intact. *Janthina*, the molluscan raft builders, are inverted snails. Instead of crawling the earth or sea floor, they blow themselves a raft of tiny bubbles and float upside down beneath the ocean surface in search of prey. These fragile voracious predators are the bluebottle's true enemies, not ferocious marlin or armoured turtles. Oblivious to the bluebottle sting, each violet snail steadily rasps its long hooked radula through whichever part of the bluebottle has the misfortune to bump into its tiny raft, slowly consuming it with relentless progress from one end to the other.

I have only ever seen the shipwrecked drifts of this living blue armada, never the fleet in full sail on the open ocean. Bernardin de Saint-Pierre 'found the sea covered with it for many days' as he sailed through the southern Atlantic.

'I have seen many of them together, ranged like a fleet of ships,' he wrote. 'From the lower part of the *galere* hang several long blue filaments, with which it seizes whatever attempts to take it. These filaments burn immediately, like the most violent caustic . . . The *galere*, while alive is the most beautiful colours; some of them are of a sky-blue, and some of a rose colour . . . I saw, in the latitude of the Azores, a kind of shell-fish, floating and living on the surface of the seas, shaped like the beard of an arrow, or beak of a bird. It is small, transparent, and very easy to break . . . Here also, we found some snails, that were blue and floated on the surface of the water, like bladders filled with air. Their shell was very thin and brittle, and filled with a liquor of a beautiful purplish blue colour.'

These floating ecosystems of pelagic organisms are called pleuston rafts, rarely noticed until they are swept ashore in massed tragedy. I do not even know if they are still as abundant as they were when the *Étoile* plunged south through the Atlantic waters towards the Pacific with Jeanne Barret on her deck, or if their numbers, like so many other species, have declined, replaced instead by great spiralling gyres of immortal plastics.

The ships stayed in the Strait of Magellan for almost a month. Many of the officers and crew had already been here on earlier expeditions, but everything was new to Jeanne and

Commerson. They watched gleaming silver dolphins with black extremities – surely the most beautiful of all the inhabitants of the sea – pirouetting in the bow wave, before effortlessly speeding ahead and disappearing into the depths. They gathered a good collection of plants, daily finding new ones. Nassau-Siegen went botanising with Jeanne and Commerson, in snow and bitter wind even at the height of the southern summer.

Covington and Darwin encountered similar challenges collected in Tierra del Fuego.

'These islands are completely forested mountains,' wrote Covington, 'their tops capt with snow which remains the whole year round. Near the summit of the mountains, there are very thick, low bushes, and patches of moss where you sink ankle deep – which makes walking very laborious. On the tops of the mountains at places where the snow has melted, you find rocks of a slaty and crumbling nature. Here, sometimes the wind blows with fierceness, which obliged us to return down to the woods, for without exaggeration we could scarcely breathe. On the mountain heights one finds plenty of guanacos, which are very shy. Their flesh is very good eating but dry. Both on the high and low woods there are great many birds of different species and by the sea, there are plenty of geese, ducks, and seals.'

When the botanists Banks and Solander were collecting here on Cook's first voyage to the Pacific in 1768, a year after Bougainville, their two servants collapsed in the snow and died of cold. Jeanne seemed at little risk of this, despite the terrible conditions. The snow lay a hand's depth on the deck on New Years Day of 1767.

'Frightful weather,' said Bougainville. 'I do believe this is the worst climate in the world.'

But the collecting continued apace, and they gathered at least 353 plant specimens. Jeanne matched Commerson and

Nassau-Siegen in her efforts, perhaps anxious to cast off any doubts that remained about her identity.

'In the Strait of Magellan, these exertions doubled, spending entire days in the forest with snow, rain and ice to seek plants or along the seashore for shells', Vivez reported. Even Vivez begrudgingly acknowledged that, 'in her praise, she generally surprised everyone by the work she did [and] the suspicions one had harboured lessened daily for lack of evidence'.

Finally, the weather started to clear and the ships tried to leave. The smaller *Étoile* had no trouble clearing the rocks and islands, but the *Boudeuse* struggled in the tight conditions and took ten days to reach clear water.

Caro says they celebrated their escape with hymn and a service – a *Te Deum*. But Commerson had an even better plan.

'We will sail into Pacific Seas,' he declared optimistically, 'in slippers and underpants.'

9

PACIFIC VOYAGERS

The Tuamotus, March 1768

It was a momentous day when we finally 'turned left' to head north up the east coast of Australia. It had taken some two and a half years since we left Port Lincoln to get this far, stopping at different ports from time to time for work. Not quite as dramatic as escaping from the frozen grip of the Horn, not quite 'slippers and underpants' weather, but still, our first entrée into Pacific waters. Almost immediately, the wind gentled and the waters warmed from the eastern current that sweeps tropical waters south down the coast. From here on the weather would only get better and the harbours safer. We stayed a night beneath the great granite lighthouse on Gabo Island (coming almost too close for comfort) and watched the penguins swim like shooting stars in the phosphorescing water. Here and there, larger shapes stirred the deeper waters, lighting up the darkness with an eerie and ominous glow.

Our first port of call was in Eden. Aptly named, the town nestles in Snug Cove, lined with white beaches and dark forested mountains. My grandparents had forwarded all our mail to the local post office, including a bundle of marked assignments and letters from school. Having just started high school, one of my new teachers had suggested that, instead of completing the designated assignments, I could swap them for a project on the area I was visiting. Eden's famous killer whale museum, with its remarkable story of cooperative hunting between humans and orcas, became my first self-directed project.

Twenty years later this small school project would transform into my first book.

Fifty-four days after leaving the Strait of Magellan, in the small dark hours of morning, the *Étoile*'s captain ordered a change of course by a single compass point. He did not say why. It was as if he had smelt land: some all but imperceptible refraction of wind, wave or atmosphere. The lookout on the *Étoile* soon sighted land from the mainmast, even as the *Boudeuse* sailed blindly ahead. Whatever Commerson thought of their manners, the Breton sailors knew the sea.

Just after sunrise, the lookout confirmed a chain of small islands to the south-east gradually emerging as the sky lightened. Before long, the clear weather allowed everyone on the poop deck to see the archipelago clustered in the distance.

Another island appeared to the west. A verdant wooded mound rimmed in a fine line of white sand. Were it not for the height of the vegetation and the tall coconut palms that sprouted above the canopy, it might have remained entirely invisible. They imagined themselves the first people to sight this island in 165 years since the Portuguese navigator Pedro Fernández de Quirós had passed this way, although there was no guarantee that this particular island was one that he had described.

As they approached, two dark and naked figures stood out against the pale sand, leaning on long spears. Fifteen or twenty more emerged from the forest edge near a cluster of small huts. The ships hove to and moved in unison along the shore. The watchers retreated, leaving the two men standing firm, raising their lances and striking them on the surface of the water. It was astonishing to imagine how people could have populated such a tiny island, not much more than a league across, and surrounded by a ring of white foaming breakers. Someone lit a fire on the northernmost point, although whether in welcome or warning was impossible to tell. There was nowhere safe to land in any case.

'Who in the devil went and placed them on a small sandbank like this one and as far from the continent as they are?' demanded Caro.

As if they could not have sailed there themselves.

It is one thing to travel across the colonised world, where the exotic is just a variant of the familiar: Spanish, Portuguese, Dutch, English or French.

'All Europeans are fellow countrymen when they meet so far from home,' Lapérouse once said.

It is quite another to encounter a people, a culture, a language, a history, entirely divergent and distinct from one's own, where the only things we share are our common human traits: hunger, thirst, fear, love, sex, humour, anger, sociality. For all our commonality, there is so much to be misunderstood.

Perhaps Jeanne did not think it so strange that there should be naked people standing on this small sandbank in the middle of the Pacific. No stranger, surely, than finding a young peasant girl from La Comelle, in a man's attire, floating past on a ship made of oak, hemp and cotton.

The island of Akiaki was formed by tip of a seamount that rises 3420 metres from the ocean floor. Like an iceberg, its bulk lies below the surface; the coral atoll itself sits barely a metre above sea level. The island is part of the largest chain of coral atolls in the world. The 80 islands that make up the Tuamotu archipelago – meaning the distant isles – extend across some 850 square kilometres of ocean.

The great oceanic expansion of Polynesian culture, in the thousand years before and after the Common Era, had been built on an unmatched technology of shipbuilding and navigational expertise. And even among Pacific Islanders, the Tuamotuans were regarded as master shipbuilders and navigators. Akiaki may have no permanent population living on it today, but then, as now, it would have been visited regularly to harvest coconuts, birds or green turtle eggs. Its closest neighbour, Vahitahi, is only 41 kilometres south-east, a daytrip for a pirogue.

These oceanic people, watching from the shore, had seen 'floating islands' drift past under their golden-foliaged trees before. The Englishman John Byron had made contact with residents in the Tuamotus just the year before Bougainville's ships sailed past. The people on the shore of Akiaki would have heard stories from their grandparents of Dutch ships, like those of Willem Schouten and Jacob Le Maire in 1615–16, or Jacob Roggeveen in 1722. Or they might have heard even older stories of the Spanish ships under Álvaro de Mendaña de Neira sailing through the Tuamotus from Peru to the Philippines in 1595.

Others soon followed in Bougainville's wake. Two years later, Cook named Akiaki atoll Thrum Island. The Spanish navigator Domingo de Bonechea sailed through the region in 1774 and the Russian Otto von Kotzebue in 1815.

As it happened, the Tuamotu islands eventually came into French possession when Commodore Abel Dupetit Thouars came to the aide of exiled Catholic missionaries and, somewhat unilaterally, seized control of Tahiti and all of its associated assets in 1838, giving rise to what is now French Polynesia.

Europeans have continued to sail past the Tuamotus, seeking to impose their own cartography on an isolated watery world that defies their understanding. Robert Louis Stevenson and his wife Fanny passed by some of the atolls on their yacht *Casco* in 1888.

'Lost in blue sea and sky,' Stevenson described the island. 'A ring of white beach, green underwood, and tossing palms, gem-like in colour; of a fairy, of a heavenly prettiness. The surf ran all around it, white as snow, and broke at one point, far to seaward, on what seems an uncharted reef. There was no smoke, no sign of man; indeed, the isle is not inhabited, only visited at intervals.'

In Jack London's *South Sea Tales*, the captain of a burning ship must relinquish control and navigation of his ship to the 'memory chart' of an old Pitcairn Islander, who guides them to safety through the shifting waters of the Tuamotus. But efforts to appreciate Polynesian expertise and knowledge of their own region invariably seemed to falter against the ironclad determinism of the European imagination. I can't help but wonder what old Polynesian navigators would have thought of Thor Heyerdahl's drifting balsa-wood raft *Kon-Tiki* bumping into Raroia to 'prove' a mythical Polynesian migration route from South America in 1947. It wasn't until the 1960s that anyone thought to ask the Polynesian navigators themselves where they came from and how they spread so effectively across their vast empire.

If the unfamiliar spectacle of land and people was not enough to bring Jeanne to the deck after so many weeks at sea, perhaps the birds would have. They filled the sky and the shore around the tiny island in 'infinite numbers'. Their wavering cries and the pungent smell of guano would have wafted across the ships, even as they stood offshore. So many birds promised rich fishing grounds. These mid-oceanic islands – so close to marine food supplies and so far from the malign influence of mammals – were clearly a paradise for birds.

But not any more. French research led by Caroline Blanvillain found that several critically endangered species – two doves, a sandpiper, a reed-warbler, a crake and a curlew – had all dramatically reduced their range across eight different islands since 1922. Polynesian rats, which first settled with humans on the islands around 300 BC, then European rats from the 1800s, are responsible for most of these localised extinctions, with cats an added burden.

In the last twenty years, sea levels have risen in the Pacific, storm surges have increased, the incidence and severity of cyclones have magnified. Can the tiny coral atolls dotted across the Pacific survive this onslaught? The answer depends on what you mean by survive. Sea-level rises average between 3 and 5 millimetres per year – well within the tolerance limits of most reef systems. Coral continues to grow upwards to maintain its proximity to sunlight. Cyclones do unprecedented damage, but within the protective ring of reef, many atolls gain sediment in storms rather than lose it. But the answer for humans is different. Humans and their associated animals and crops are not so tolerant of saltwater incursions. And human activity, removing vegetation and altering shorelines, seems to dramatically increase the damage done by weather.

I met a woman from Samoa at a friend's wedding about fifteen years ago. When I asked her which island she was from she laughed and said I would not know it. She explained to me that at high tide, her home island was almost completely submerged now. The pigs, which spent most of their time foraging for crustaceans on the reef, were all but aquatic. The people too relied mainly on the sea for food – their only transport was by boat and their houses had to be built higher and higher to keep free of the encroaching waters. It was getting harder and harder to grow taro in the salty soil and only the invincible coconuts provided a reliable source of vegetable. Her home island was a day's boat trip from the nearest island with a radio, a week away from anywhere with a runway for small planes. She was right – it was almost impossible for me to even conceive of people living in such an isolated and vulnerable place. And I wonder now, so many years later, if anyone is still living there.

I am surprised to realise that the sea does not rise evenly across the earth. Strangely, it rises faster in some areas than in others. In the Solomon Islands the sea is rising by an average of 7–10 millimetres per year, three times faster than the global average. In the north of the country, the estimates are 16.8 millimetres per year. In 2016, five reef islands were lost in the Solomons. Nuatambu Island, home to 25 families, lost half of its habitable area and eleven houses have been washed away since 2011. Many villages have been relocated inland, onto higher ground, although this is not an option for coral islands. Even the capital of one province, Taro, is set to relocate its 600 residents, four churches, hospital, school, police station and courthouses to a larger island. The former market site is currently chest-deep under water. They have already been evacuated three times after the island was swamped by tsunamis triggered by earthquakes.

In the 1960s the French finally found a use for their far-flung colonial outpost on the other side of the world, and detonated mushroom clouds over Mururoa and Fangataufa atolls, some 400 kilometres to the south of Akiaki and 1200 kilometres south-east of Tahiti, showering plutonium across 3000 kilometres. The impact of plutonium in the Pacific environment – not just from French testing but also from US testing in the Marshall Islands, UK testing in Australia and Soviet dumping of radio-active liquid waste in the north Pacific – has barely even been examined. Mururoa and Fangataufa remain off limits to the public. Leukaemia and cancer have become major causes of mortality among Tahitian workers exposed during the testing. But no independent monitoring of environmental impacts has been permitted by the respective authorities.

The environmental future of these far distant islands, so rich and idyllic when Jeanne first saw them and so remote from the worst excesses of human activity, looks surprisingly and unfairly bleak.

My parents owned a little brass plaque when I was small, glued to a cheap black frame that sat on our bookshelves. When we moved onto the boat, Dad fixed it to a block of wood and attached it to the bulkhead.

I don't recall the picture on it exactly. I'm sure there were distant mountains, trees and probably someone with a bundle tied to a stick, looking longingly into the distance. But I remember the poem, long after the plaque itself was lost.

Don't feel afraid of anything
Through life just freely roam
The world belongs to all of us
So make yourself at home.

It was a twee little piece of kitsch, but an apt motto for 'boaties' who sought to free themselves from the constraints and regulations of suburban life. You can sail, more or less, where you want on the open ocean. You can come and go as you please, with only the weather to dictate terms. Harbourmasters and authorities rarely pay you much attention, beyond directing you to the nearest wharf or suitable mooring point or, in more remote areas, swinging down in aircraft to check for smuggling.

That little poem gave me a great sense of confidence, as a child, that I could find a home, or make one, wherever I went. That home was something you took with you, not something you left behind. It was a notion that echoed in my mind every time I launched my kayak or rigged the dinghy in a new port and set off to explore the secret creeks and hidden bays, with my cat standing in the bow oblivious to the risk of falling.

But I cannot escape the fact that it is also a perfect motto for a colonising culture, for those who sought to make new homes in new lands across the seas. We expect to be made welcome wherever we go. And I can't help but see myself in the ignorant and arrogant responses of those explorers who so often refused to accept that, most of the time, they were quite simply very clearly unwelcome.

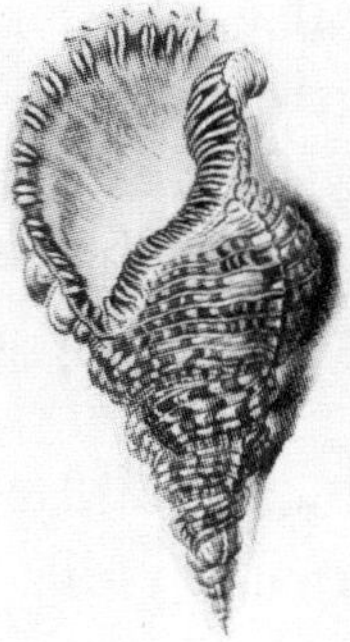

10

REVELATION

Tahiti, April 1768

To STAND ON THE edge of a volcanic island as long tongues of lava flow inexorably to the shore, to see the red-hot fissures spit and hiss as they strike the cold sea, to watch the light of primordial rock fade as it freezes mid-motion, forged unwilling into fierce solidity, is to bear witness to the world's creation.

Amid the sulfuric stench and heat, a beetle blows ashore on a leeward breeze, catches a foothold and clings, before crawling off into the darkness. On the black sands of the beach, a pile of driftwood washes ashore, laden with seedpods and debris from other lands. A white gull inspects the booty with an unblinking eye in the dying light.

This bleak and sterile birthplace has generated a panorama of islands that illuminate the origins of life on land: the endless progression of ecological establishment – succession, radiation and speciation. Each island in the chain, shifting with the moving plates away from the crack in the earth's crust that created it, represents a step in the process from raw, bald, rocky infancy to the lush, verdant maturity of the oldest, most distant isles.

Tahiti is part of just such an island chain. It is one of the Society Islands that started erupting from a volcanic hotspot in the middle of the Pacific plate some 4 kilometres below the ocean surface 4.5 million years ago, completing its construction with Maupiti less than a million years ago. Like all Polynesian islands, like all landmasses ultimately, Tahiti emerged from the molten depths of the earth's mantle to create the terrestrial plane on which we walk, on which our very existence depends.

The terrestrial species that live in Tahiti have come from far away, diversified and speciated, into a range of distinctive yet related forms. In a process common to all oceanic islands, a few hardy colonisers settled, occupied and took up new niches. They separated across valleys and mountains, across habitats and ecosystems, and evolved new forms different from each other. Plants arrived first, courtesy of the specialised survival and dispersal pods in which they often package their seeds, or their capacity to take root and regrow from broken fragments of themselves washed ashore. Insects, perhaps, arrived next, lightweight and aerial, capable of surviving even the freezing turbulence of the stratosphere. Eventually a few birds too, found their way across 4000 kilometres of ocean. Even terrestrial reptiles swept ashore, thanks to their slow metabolism and resilience. Lizards (and on some islands even giant tortoises) arrived via sea drift. A bat or two might find their way here, but other than the marine mammals, it is rare to find any terrestrial mammal or amphibian native to an oceanic island.

It is the unique combination of their isolation and sparse seeding with original species that seems to supercharge evolution, resulting in intense diversity in a remarkably small area. Of the 467 species of plants found on the island of Tahiti, 212 are found nowhere else. When the first Polynesians approached these islands in their large ocean-going canoes in 300 CE, they would have known they were approaching a rich and fertile land, perhaps even before the mountain peaks had broken the horizon. They would had known from the cries of the seabirds wheeling overhead, from the subtle shift in the currents, from coconuts and palm fronds floating offshore, and from the shifting seasonal, geographic and vertical distributions of fish species. And soon enough, the damp rich scent of fecund forest and foreshore would have confirmed their hopes and drawn them ashore to a land of perfect weather, abundant food and easy living.

When the French ships arrived here 1500 years later, after two months at sea, they too would have quickly realised they had arrived in a paradise for sailors and biologists alike.

'Scarcely did we believe our eyes,' wrote Bougainville, 'when we discovered a peak laden with trees up to its isolated summit . . . it would have been taken from afar for a pyramid of immense height, which the hand of a skilful decorator had adorned with garlands of foliage . . . The less elevated land is interspersed with meadows and groves, and all along the coast, at the foot of the high country, there is a low level margin covered with plantations. It was there that among the banana trees, coconut palms and other fruit-bearing trees, we could see the houses of the islanders.'

At least here Jeanne would not have to slog through ice and snow. Here, she could expect to collect in ease and comfort, with a bevy of laughing women and children to help her, and know that every second species she picked up would be new to science.

~

Voyages are always demarcated by their destinations, by landfall. No-one wants to hear about endless days at sea, fair winds and good weather. They will listen to stories of near disaster or adversity but otherwise it is only the ports of call that are of interest.

I recount my own childhood travels, destination by destination as we daytripped our way north from Eden to Bermagui to Batemans Bay to Ulladulla and Jervis Bay. I collected fossil sea urchins from shaly cliffs and admired the bleeding tooth nerites that replaced the plain black periwinkles from home. A large orange seahorse that washed ashore quickly faded in the alcohol I tried to preserve it in. Clouds of blue soldier crabs swept in waves across the mudflats at our approach. Some of my collections grew legs and ran away. Hermit crabs had claimed prior residency to my treasures and regretfully I returned them all to the beach, simultaneously enchanted and disappointed by this new obstruction.

The trip to Sydney was an overnighter from the serene and beautiful Jervis Bay. Behind the cliffs and cloak of dark vegetation the city was barely visible but for the great halo of light that warned us hours ahead of civilisation's approach. I wrote a terrible poem about it for an English assignment, which was duly published in the school magazine. I favoured poetry over prose at the time because it was shorter and I was lazy. As I struggle now to recall the details of a journey so long ago, I wish I'd written more.

As we rounded Barrenjoey Head, Broken Bay opened north to Gosford Waters, south into the Pittwater, and west into Cowan Creek, Berowra Waters and the sinuous Hawkesbury River, all encased in sandstone cliffs and eucalypts. These are some of the most beautiful and most spectacular natural waterways in the world. And, aside from the mansions lining the north shore and the armada of yachts at anchor, you might never even know you were on the edge of Australia's largest city.

The diaries from the Bougainville expedition are full of hastily sketched profiles of Tahiti's mountainous islands, drawn from the deck of the ships. I could tell you what the Frenchmen saw, what they thought, and what they imagined as they approached. But I cannot tell you what Jeanne saw or thought or felt as she approached the moment of her dénouement.

Tahiti is not just a turning point in Jeanne's story and hers is not the only voice silenced here. We hardly ever hear what the Tahitians thought as these European ships approached, carrying their weapons of violence, their plagues of contagion and their seeds of religious and political corruption. The first woman to sail around the world is as obscured here as the first Tahitian to discover Europe. Ahutoru is central to Jeanne's story. This is his story too.

As with Jeanne's story, I have to pull the strands of Ahutoru's narrative from many different sources. Bougainville tells me one thing, Vivez another and Caro still more. Microscopic fibres carefully separated and woven back together into a delicate filamentous fabric that is barely a shadow of its original form.

Word had travelled fast on the islands of the approaching canoes. They were longer and higher than the largest war canoe. Just one of them could conceal a raiding party of 100 men. The trees that grew from their decks were so large that men clambered through them like bats in the canopy.

This time there were two – even from the shore Ahutoru would have seen that these vessels, these people, were different from the last ones to visit Tahiti.

The ships tacked back and forth up the coast. For a while it looked as if they might sail past, back into Matavai Bay near the marai of Purea where the last canoe had landed. But they doubled back, coming in at Hitia'a, near the marai of Ahutoru's own chief and Purea's rival, Reti. Ahutoru had no time to waste getting on board.

No-one knew where these strangers had sailed from or what they were doing. They did not really seem to know themselves. They had no knowledge of where they were or where they were going. But they had weapons more powerful than anyone had ever seen, as the blood spilt in Matavai Bay last year had attested. After a brief but decisive battle, Purea had decided to make allies of her unwelcome visitors. Reti had been unimpressed by this enhancement of his rival's status. This time, he wanted his own allies.

Ahutoru, as Reti's envoy, set out to intercept the ships, keen to make the acquaintance of these strange men. And, more importantly in Ahutoru's opinion, make the acquaintance of any new women.

There was a good reason for Tahiti to be known as Aphrodite's Island. Bougainville called it New Cythera, after the Greek island devoted to the goddess of love. Every journal from the voyage was suddenly filled with woman, young or old, nubile or unattractive, all offering sex, willingly, unwillingly, boldly or shyly. It all depends on the telling. These women have no names and no voices. Only their actions and bodies are described.

If Bougainville had landed in Matavai Bay, where the English explorer Samuel Wallis had arrived a year earlier in the *Dolphin*, he might have had a different view of the women. Wallis dealt exclusively with Purea (or Oberea), whom he described as the Queen of Tahiti. When Cook visited in 1769, he too met Purea, whom he described as 'head or Chief of her own Family or Tribe but to all appearance has no authority over the rest of the inhabitants whatever she might have had when the *Dolphin* was here'. Cook described her as 'masculine', while his botanist Joseph Banks described her as 'lusty'.

Bougainville did not meet with Purea during his visit. He dealt exclusively with Reti, a neighbouring chief. He talked only to the men. When he mentioned their wives, his descriptions were clear and precise – as if these were women he could understand: women under the authority of their husbands, required to do as they were told. The younger unmarried women, who had made their presence felt almost as soon as the ships arrived, confounded him. He struggled to find words to explain women who appeared to be under no-one's control. His language became flowery and romantic, referencing nymphs and classical goddesses, as if he could not find a model to describe these women who seemed to operate with complete sexual freedom.

'A young and fine-looking girl came in one of the canoes, almost naked,' commented Bougainville prosaically in his journal on the first day, 'who showed her vulva in exchange for small nails.'

By the time he was writing up his account for publication, however, he had refined his literary style.

'In spite of all our precautions, a young girl came on board, and placed herself upon the quarterdeck, near one of the hatchways, which was open, in order to give air to those who were heaving at the capstern below it. The girl carelessly dropt a cloth which covered her, and appeared to the eyes of the all beholders such as Venus shewed herself to the Phrygian shepherd having, indeed, the celestial form of that goddess. Both sailors and soldiers endeavoured to come to the hatchway and the capstern was never hove with more alacrity than on this occasion.'

There was much in Tahitian customs that the Europeans did not, could not, or would not understand. Sex and property, the twisted shackle at the heart of nuptial relations, were equally incomprehensible.

'The king is the first and greatest of thieves,' Caro observed of the Tahitians, entirely without irony, even as his own commander demanded free wood, water and food, without the slightest knowledge of the currency of exchange. The laws of *tapu* and *muru* and their variants, which reigned over Polynesian communities, were invisible and inconceivable to the Europeans. No wonder men like Cook and Marion du Fresne fell fatally foul of them. It has taken decades for anthropologists to try to reach the most basic of understandings of these systems, scurrying to document them as fast as religious and secular forces erased them from collective memories.

Frederick Edward Maning, known as a 'Pākehā Māori', was Irish-born but chose to live and marry into the Hokianga Māori community. His recollections in *Old New Zealand* illustrate that to be robbed, to be thought worthy of being robbed, was both a privilege and an honour. What the European's saw as theft, the Polynesians saw as a vital redistribution of wealth through the community: a social and economic necessity. Thousands of pages of text have been devoted to concepts of property – the spirit of the gift – the obligations generated by gift-giving and the consequences of failing to meet those obligations.

I could not hope to explain any of the complexities of either this field of research or the cultures it seeks to explain. I am as blinded by my own cultural and historical blinkers as any of the early explorers. My inability to fathom Tahitian attitudes towards death, violence and sexuality matches my inability to understand eighteenth-century French attitudes towards – well – death, violence and sexuality. Cannibalism or Catholicism, *tapu* or the divine rights of kings. It's all a foreign country to me.

I suppose the best we can do is to recognise that our blinkers are our own impediment, and that just because we can't see something, doesn't mean it isn't there.

I don't really understand why Commerson spent so much time writing about the people he visited and so little time describing the natural history that was supposed to be his passion. He seemed to write more about the Patagonians and Fuegians than he did about the rich marine life of the Magellan Straits. He went into graphic detail about the lives of the Tahitians while barely mentioning the extraordinary coral reefs or the distinctive ecology of these volcanic islands. I don't understand his fixation on humans when there was so much that was new and unknown around him.

I started by studying psychology and anthropology but quickly slid sideways into animal behaviour then zoology, once I realised that animals were much more interesting and diverse than humans. Commerson, at the very beginning of modern biology and anthropology, had a different view. The study of humans – more particularly 'other' humans – was natural history, and was at the forefront of Commerson's mind as a naturalist traveller. When the Duc de Praslin asked him to advise on the study of natural history for future naturalist travellers, Commerson put the observations of humans first.

'What is there in fact more essential to observe in any country you enter for the first time than the races of men who inhabit it, their faces, their manners, their people, their clothing and their weapons?' Commerson asked.

It's also an artefact of the way he wrote science. He was, like most naturalists of the time, a taxonomist. He was interested in discovering, describing, labelling, naming and claiming. Biology had yet to develop an interest in the behaviour, ecology, environment or biogeography of species. Commerson confined his descriptions to his notebooks, organised by class and species and written in Latin. Hardly anyone seems to have even read them.

I shouldn't be disappointed. And I can't really blame him. After all, I thought there would be a lot more natural history in this book, too, but instead, I find myself writing more about the people.

I can find only one species collected from Tahiti by Jeanne and Commerson. It's a fish – a birdnose wrasse. A strange little creature, it has a distinctive long pointed snout and a jerky flapping motion when it swims, as if it was indeed masquerading as a bird. In life they are an unvarying shade of blue, save for a tint on their fins and a beautiful emerald and aquamarine eye. But they lose their fine colours very quickly in death, and after 250 years in a museum collection the three specimens that Jeanne and Commerson collected are desiccated, shrivelled and yellow, best recognised from the tag on their tails.

These fish are hard to identify when young, before they have developed their characteristically long nose. Like many wrasse, they are hermaphrodites. They begin life as small, rather dull, females but in older age develop brighter colours and turn male. In some other wrasse species, the females can change sex and function as males at various times.

Commerson would not have been aware of the wrasses' interesting sex-changing habits at the time. And he did not think it looked like a bird either. Instead, he described its snout as being like a *clou* – a nail.

Nails were on everyone's mind. The Tahitians might have been impressed by the European's guns and axes, but nails revolutionised canoe-building. Even the smallest of nails could be traded for sexual favours. And clearly even the biologist could not stop thinking about them.

Perhaps it was no great surprise that Commerson was distracted from his focus on natural history while he was in Tahiti, where human activities were so diverting.

'The act of creating a fellow human being is a religious one,' he observed, 'the preludes to which are encouraged by the vows and the songs of all the assembled people, and the climax celebrated by universal applause.'

It seems likely this went beyond mere observation.

'Every stranger is admitted to share these delightful mysteries; it is even a duty of hospitality to invite them to do so. So the good Utopian ceaselessly thrills at either the sensations of his own pleasures or the spectacle of those of others.'

Nassau-Siegen left no doubt about his participation either.

'I was strolling in a charming place,' he recounts, 'carpets of greenery, pleasant groves, the gentle murmur of streams inspired love in this delicious spot. I was caught there by the rain. I sheltered in a small house where I found six of the prettiest girls in the locality. They welcomed me with all the gentleness this charming sex can display. Each one removed her clothing, an adornment which is bothersome for pleasure, and, spreading all their charms, showed me in detail the gracefulness and contours of the most perfect bodies. They also removed my clothing. The whiteness of a European body delighted them. They hastened to see whether I was made like the locals and pleasure quickened this research. Many were the kisses, many the tender caresses I received.'

I sometimes wonder if it is ever possible for anyone within a Christianised society to comprehend a world of sexual freedom or promiscuity, in whatever form it takes. Particularly through the even more recent lens of Reformation, then Victorian, prudery. Christianity came fast and fierce to Polynesian society, exposing them to their nakedness and bringing shame to the Garden of Eden. This conversion has closed a door to the past, making it almost impossible to understand what has been lost.

As George Orwell once said, 'the most effective way to destroy people is to deny and obliterate their own understanding of their history.'

For the European observers, the Tahitian women (like their own) were invariably discussed in terms of trade and commerce. They were given by men, or were trading or bartering their services in return for nails. I have read papers in which every sexual encounter in Tahiti is described in terms of rape. Others use the language of coercion or prostitution. Few are willing to consider that maybe these young women were, like Ahutoru, just interested in the sex. Perhaps there wasn't anything more to it than that.

The Tahitians must have been even more confused about the sexuality of their visitors. For one thing, at first sight, it seemed almost impossible to tell whether they were male or female or something else entirely. The Europeans all looked much alike – small, thin and pale, with no beards, unadorned by the tattoos and decorations of either men or women or status, and covered in bland indiscriminate clothing that concealed their genitals. Nor did any division of labour distinguish them – they sewed, cooked, cleaned, hunted and swept – performing men's as well as women's tasks. They were sexually ambiguous.

And what kind of people travelled long distances in a boat

without bringing any women with them? Apart from a raiding party. No wonder Indigenous people throughout the Pacific were puzzled and concerned by these strange visitors.

The travel narratives are full of stories of sailors having to strip naked to prove their manhood. Bougainville's cook discovered this when he slipped ashore against his commander's direct orders.

'He had hardly set his feet on shore,' reported Bougainville, 'when he was immediately surrounded by a crowd of Indians, who undressed him from head to foot. He thought he was utterly lost, not knowing where the exclamations of those people would end, who were tumultuously examining every part of his body.'

Bougainville's cook thought he was going to be eaten, but his clothes, if not his dignity, were soon restored. He was offered the opportunity to demonstrate his manhood with a young girl, but he was in no state to achieve that which he had so ardently desired when he snuck ashore and, with all good will, he was taken back to the ship to beg his commander's forgiveness.

The Tahitians soon realised there were plenty of men aboard the ships, but they still wondered where all the women were hidden.

Ahutoru was the first Tahitian to board the *Étoile*. He would also become the first Tahitian to discover Europe, when he returned with Bougainville to France. A native of Raiatea originally, and exiled from his home island by Bora Bora warriors, he was about 30 years old when the French ships arrived. Yet we know relatively little about this great adventurer and explorer. There are no pictures of him, little information about his family, his upbringing, his life or his thoughts. He was variously described as larger and stronger (than the French), smaller than other

Tahitians, or of mediocre size; as both black or 'mulatto' in complexion, as attractive and handsome or ugly and stunted, as highly intelligent and articulate or incredibly stupid with a speech or mental defect. There is not much to go on beyond long black hair, a beard and tattooed hands and thighs. Like Jeanne, he is an anomaly, a diversion to the main story of European discovery, worthy only of a passing paragraph here and there, a humorous anecdote or dismissive comment. But he is also central to Jeanne's story.

Ahutoru came alongside the *Étoile* in a large canoe, which effortlessly kept pace with a ship making a good few knots under sail. An imposing figure, almost Romanesque in a sleeveless white shift, he gestured imperiously that he wished to board. The crew threw him a rope, which he grabbed with both hands, wedging his feet in the bow of his canoe as it surged against the pull of the ship. The rope stretched perilously taut under the strain. The crew shouted for him to let go, but Reti's envoy could hardly fail a show of strength. Ahutoru tightened his grip and hauled himself in, hand-over-hand.

He seized the rigging chains in one fist and swung himself aboard, leaving his retinue to trail in the ship's wake. He strode the deck as if he owned it, dwarfing the Frenchman not so much in height as in strength and stature. It was as well he came armed only with a peaceable expression, a green branch and a ready smile. He was here to charm, as well as impress.

Ahutoru invited the French ashore, promising an abundance of food and drink while he inspected every aspect of the ship and crew. In a show of good will, they exchanged his gift of *tapa* cloth for European clothes, although it was hard to find any big enough to fit. Despite the captain's protestations, Ahutoru waved his vessels away, deciding to stay the night and dine with his visitors, as oblivious to his host's discomfort as the French would be to the Tahitian's concerns on their landing.

Both Vivez and Caro describe how Ahutoru laughed, drank and ate in the mess, his keen interest in women and how he watched the crowd who gathered around him. Everyone told him there were no *ayenne*, no women, on board, but he clearly did not believe them. They would be standing back perhaps, watching from the sidelines. According to Vivez, Ahutoru spotted Jeanne and Commerson, standing near Labare, the armourer, at the back.

He leapt up, pointing, '*Ayenne*!'

Everyone looked at Labare, a delicate young man who might easily be mistaken for a girl. But the Tahitian was not looking at Labare. He singled out Jeanne.

If Vivez was delighted by this turn of events, Commerson was aghast and Jeanne fled.

Ahutoru was left to puzzle over these strange creatures who concealed their women and shot rockets into the night sky like angry spirits for amusement. He slept on the deck and followed the ship's route by the stars as it tacked back and forth off shore, so that at least he would know where they were going, even if his hosts did not.

The next day the two ships managed to find a space between the reef and the shore where they could anchor safely and load water. The commander ordered the sick ashore, but no sooner had they landed and begun erecting their tents than Reti and Reti's father arrived, directing them to load their luggage back into their boats.

The French were no more welcome to sleep ashore than Ahutoru had been on the *Étoile*.

There was a careful discussion between the commander and Reti. Bougainville explained with eighteen stones the number of nights they planned to stay. Reti countered with nine stones.

An agreement was reached. The tents were erected and both Reti and Bougainville, along with their respective entourages, spent a wary night in each other's company.

While the crew loaded water the next morning and others traded nails for coconuts, bananas, fish, crabs and crayfish, Commerson and Jeanne came ashore, eager to do some collecting.

They were met by an enthusiastic crowd of Tahitian men, shouting '*Ayenne*! *Ayenne*!'

There was no doubting their intentions. It was obvious that, like Ahutoru, they thought she was a woman. And it was equally obvious that they were offering her the same attention and hospitality that the young Tahitian women had been offering the French men.

One of the Tahitians picked up Jeanne and swung her easily over his shoulder, intending to carry her away. The French officer on guard swifly drew his sword and intervened. Jeanne hurried back to the longboat and returned to the ship.

But everyone who was ashore had seen it. And before long everyone who was aboard the ships soon heard. There could be no doubting it now. Every suspicion, every rumour and every tattle of gossip that had smouldered over the last year, now flared and exploded into life. The Tahitians had seen through Jeanne's disguise. No-one would believe her denials and her explanations any more. Her cover had been catastrophically blown.

Jeanne Barret was a woman.

There was no reprieve for Jeanne on the ship. Ahutoru paid court assiduously and persistently. Now Commerson could not go ashore either for fear of leaving Jeanne alone with this new admirer. Ahutoru plied them both with questions that they did not really understand and they answered with words whose meanings they did not know. Perhaps Commerson tried to explain, as he had done before, that Jeanne was a eunuch. However it came about, somewhere in the conversation *māhū* was mentioned – a Tahitian word for one who combines both genders.

According to Vivez, Ahutoru's attitude changed immediately. He remained attentive and appreciated being combed, powdered or dressed by Jeanne, a very common activity among all the sailors, but he ceased pursuing her. He remained respectfully friendly until the ships left Tahiti, when he transferred to the *Boudeuse*, to continue the voyage with his new friends.

Ahutoru's description of Jeanne as *māhū* must be the first record of the Tahitian term for transgender people in the European literature. This word refers to someone with the physical characteristic of one gender but who lives and identifies as the other gender. It is something people are born with, not something they choose, and *māhū* occupy a respected and acknowledged place in Polynesian societies.

In the historical and contemporary accounts, though, *māhū* are almost exclusively described as males who dress and behave as women. In English at the time they would be termed transvestites or in French, *travesti*. But this does not seem to reflect the way Tahitians see the term.

'*Māhū*,' explained a Tahitian woman to a researcher, 'that can be a man or woman because that's what it means, someone who's both.'

Māhū is the Tahitian and Hawai'ian term for what Samoans and Tongans call *fa'atama* or *fafine tangata* (to become female) and *fa'afafine* or *fakafefine* (to become male). The terms refer to someone's gender, not specifically their sexuality, but the sexuality of *māhū* seems to follow the traditional heterosexual preferences of the gender they have adopted.

The French were amazed that Ahutoru and the other Tahitian men could so readily identify a woman who had convinced them for so long that she was a man. Charles-Marie de la Condamine would later implausibly argue that Ahutoru had 'a sense of smell so exquisite as to distinguish . . . the difference between the sexes'. Besides, the question was not how the Tahitians knew that Jeanne was a woman, but how Jeanne had managed to convince the Frenchmen that she was not. As Bougainville would ask, 'how could we discover the woman in the indefatigable Baret' when she did not behave at all like a woman? Appearances are everything. If a woman knows herself by her coat, as the French saying goes, but she refused to wear the right coat, how was anyone to know who she was?

For much of the journal that Duclos-Guyot and Commerson shared, it was the young Pierre who wrote the main entry and Commerson who added marginalia or an occasional postscript. But on 6 April, shortly after their arrival in Tahiti, Duclos-Guyot stopped journal entries and for the next week the only entries were made by Commerson. Duclos-Guyot had barely missed an entry in the entire voyage. I can only assume he was ashore for most of the stay in Tahiti. The day Commerson took over the journal was the day Jeanne's identity was revealed on shore. He does not mention it.

Commerson wrote about the ships being surrounded by canoes on the first few days. He described the large handsome Tahitians – 'affable, nude and unarmed' – with long black beards and beautiful teeth, inviting themselves on board. He described their gifts of coconuts and bananas and the prodigious fires that burnt all along the shore at night.

But he said nothing about going ashore on the 6th. There are no entries for the next three days. I assume he went ashore on the 10th because he says that 700–800 'Indiens', both men and woman, crowded around their tents to admire them.

It was not until they had been in Tahiti for a week that Bougainville noted, 'Mr de Commerson began making botanical specimens this afternoon'. The Tahitian women and children vied with each other to bring him bundles of the plants and buckets full of assorted shells. Limpets, mud creepers, oysters, scallops, cockles, cones and cowries. A giant trumpet shell used in ceremonies. Turban shells and trochuses that hid their pearlescent nacre beneath bland and unassuming attire. Surely many of these must have made their way into the collections on the ship.

On the 12th and 13th, Commerson discussed a dispute in which three or four Tahitians were killed and described Bougainville's subsequent efforts to make peace. It sounds like Commerson was ashore then too. On the 14th, they had all returned to their ships and prepared to leave.

All up, I can account for just four days when Commerson was ashore on Tahiti. It doesn't leave a lot of time for botanising. For this period, while Jeanne was stuck on the ship, I cannot find a single Tahitian specimen represented in any of the herbaria records from this voyage. Her absence was a costly one.

Ahutoru boarded the *Boudeuse* on 15 April. Caro said that the islander's main motivation was his interest in European women. If that were true he could have stopped in Jakarta or Mauritius, but he sailed on to France, where he remained for a year as Bougainville's guest. Ahutoru himself tells us nothing of his intentions – he never learnt to speak French for all he nonetheless managed to communicate extremely well, charming and navigating his way successfully through all levels of French society from the King and the Duchess of Choiseul to the dancers and singers at the opera.

George Forster, the pompous yet brilliant linguist who sailed with Cook on the *Resolution*, and who translated Bougainville's narrative into English, regarded Ahutoru as 'one of the most stupid of fellows' of 'very modest parts', even though Forster does not appear to have met him. Others criticised Ahutoru for his failure to speak French. Perhaps Ahutoru thought the same of his hosts, whose proficiency in his language remains unknown.

Ahutoru certainly had a good knowledge of navigation by the sun and constellations, was aware of the recurrent nature of comets, as distinct from shooting stars. He knew of the location and distances of the surrounding islands up to, as far as Bougainville could tell, fifteen days sailing.

For all he did not seem able to pronounce French consonants or nasal vowels (an inability with which I have much sympathy), Ahutoru was fluent in his own language. On notable occasions during the voyage he would extemporise in long rhythmic stanzas which seemed to be recording the events in his memory. Similarly he often recited a long prayer – the prayer of kings. Perhaps it was his role as a storyteller or performer that gave him such a love of the opera and ballet when he was in Paris.

I cannot help but wonder what was contained in Ahutoru's account of the voyage. Just like Jeanne, his history has been lost.

That the Society Islanders, or the Raiateans to be precise, were great travellers was evidenced by more than just Ahutoru's voyage in 1767. Two years later, in 1769, another Raiatean, Tupaia – a priest and senior adviser to the female chieftain Purea – travelled with Cook on the *Endeavour*. Tupaia was a hugely influential figure, particularly in New Zealand, where his language skills and sharing of cultural knowledge greatly assisted Cook to successfully navigate tricky waters. From Tupaia's perspective, his main interest seems to have been in reconnecting with this Māori outpost of Polynesian society that had long been isolated. A few years later, in 1773, Mai (often known as Omai), also exiled from Raiatea, left Huahine with Tobias Furneaux on the *Adventurer*, arriving in London in 1774, and was immortalised in a painting by Joshua Reynolds. He returned with Cook's third voyage to his home island in 1777, the first to complete the journey.

Even from Wallis and Bougainville's first brief visits to Tahiti, it was apparent that the islanders recognised immediately that a great change had arrived, and they took every opportunity to learn and adapt to that change. For the next decade, the Society Islands were wracked by warfare as local chiefs, male and female, sought to align their control with this new external force. In the end, though, the explorers themselves were of the least concern for the islanders. For next came the whalers, and the traders, and the missionaries, and all of them brought devastating diseases, and life in Tahiti would be irreversibly transfigured.

11

CONFESSION

Samoa and Vanuatu, May 1768

I DRIFT ON THE surface of the water, warm sun on my back and the reef stretching kaleidoscopically beneath. An air-breather immersed by virtue of a thin plastic tube. I like the quiet of snorkelling, the only sound that of the water slopping about my head, my own wheezy breath and the distant pop and crackle of the underworld. If I used scuba I would have more freedom, could dive down and investigate the darting fish and dark crevices. But I have no wish to poke or touch, no need to catch or hunt, to investigate or pry, with the rattle of compressed gases hissing and boiling in my ears, metal-strapped to sink onto a world where I don't belong. I prefer to float in silence on the surface, in the liminal space between worlds, half air, half water, and observe.

I remember the first time I saw a coral reef – little more than an unexpected brown tint in the water instead of green or blue. *Caliph* was anchored off Great Keppel Island. It was too late in the afternoon to go swimming, and too cold, so I put on a pair of goggles and hung over the edge of the dinghy with my face in the water, watching this new wonderland drift beneath me until my limbs grew numb.

As we travelled north, the reefs grew larger and more interconnected. Tiny village outcrops gave away to great sprawling high-rise metropolises teeming with fish, coral, sponges and molluscs. It is impossible to describe the diversity, complexity and variability of the Reef – the beauty and the terror. Exploring gardens of intricate delight tended by colourful darting fish that sweep the paths and collect the litter. Or lazing among great schools of silver, within finger's touch of shimmering fearless piscine companions in the neutral territory of marine reserves.

I remember the clouds of colour sweeping in murmurations across valleys and plains. Delicate fans and fingers feathered in the currents to touch and taste. Cephalopod ghosts morphed invisibly against sand, reef and sea, before streaking away in a jet stream. A crack opened in the grey rock to reveal the black velvet eye of a giant clam, flecked glowing green and blue, like a portal to another dimension.

But there were other staring eyes and sharp teeth lurking in dark spaces. A serpentine slither or spider-like scuttle sparked primitive mammalian fears. Knives, blades, harpoons, poisons and potent toxins, concealed in a fish that looked like a rock or within the exquisite beauty of a deadly *géographe* cone.

And then, across a ridge, I found myself drifting over great plains of white bones that even then, 35 years ago, had been laid waste by a blast of coral bleaching or the depredations of marauding armies of crown-of-thorns starfish. I couldn't help but notice in such laid-waste waters, and outside the protective boundaries of reserves, the absence of giant fish that watched impassive and unassailable. Instead there was just an overabundance of the small and inedible. Grey and brown where there should be colour, a lone cloud of violet swirling pale against the blue, the great clam eyes closed or emptied, their white carapaces stolen to delineate the paths of earth-bound feet.

You cannot weep underwater or you will drown in your salty tears. So instead I tumbled across the threshold of the deep, felt the vertiginous wall drop beneath me, snatching away my breath and heartbeat as the dark shapes of an uncertain future loomed and swirled in the distant open ocean.

The *Boudeuse* and the *Étoile* sailed west into unknown waters. Bougainville's charts may have been empty, but Ahutoru knew exactly where they were. He tried desperately to persuade Bougainville to stop over at the island of Raiatea as they sailed past. He seized the helm of the ship and tried to change its course. But Bougainville refused, even though it was barely out of their way at all. Ahutoru climbed the rigging and looked longingly in the direction of his homeland as the ships sailed on.

When he finally came down to the deck, he informed Bougainville that if there were no women where they are going, they may as well cut off his head. He talked endlessly of women. Bougainville observed that this was 'his only thought, or at least all the others he has relate to this one'.

Bougainville had other concerns as they left Tahiti. Surely he would have to deal with the issue of Jeanne now that the Tahitians had identified her as a woman. He must have known about the revelation on the beach. Gossip would have been rife on the ships. Wouldn't Ahutoru have told him that Jeanne was a woman dressed as a man? Maybe. Maybe not. Perhaps an expedition commander with a few hundred men under his command

and their lives in his safekeeping had better things to do than worry about an otherwise impeccably behaved botanist's assistant who caused no trouble other than being a woman. Perhaps an experienced commander would know that it is best to let such storms-in-teacups subside on their own, that it is better to wait and see than act precipitously, particularly when there is nothing that can be done about them in the middle of the Pacific.

On the day of their departure from Tahiti, the *Boudeuse* unexpectedly hauled to and sent its boat over to the *Étoile* to collect Commerson with the message that his medical opinion was needed.

'I hurt my leg while going,' complained Commerson, somehow blaming Vivez for the mishap.

Apparently Bougainville had a fever. If he did, he recovered very rapidly. There is no mention of their conversation in either Bougainville's or Commerson's journals. I find it hard to believe that this private medical consultation would not have included a discreet conversation about Commerson's assistant turning out to be a woman.

But nothing happened. The ships simply continued on their course.

They proceeded with caution, aware that somewhere ahead lay the uncharted east coast of 'New Holland'. Jacques-Nicolas Bellin's *Carte réduite des terres Australes*, updated by Didier Robert de Vaugondy in 1760, broadly outlined three-quarters of the coast of Australia, north, west and south, but from the eastern end of the Great Australian Bight right up to the top of Cape York was largely one long *côte conjecturée*. The question was not whether they would be able to find the Australian east coast, but whether they would be in a state fit enough to land and explore it.

There was nothing to relieve the long days on the ship. In the tropical heat, the smallest irritation of salt-washed clothes scratched and abraded the skin, opening wounds that soon oozed with infection. They drifted by enticing islands of thickly wooded slopes filled with botanical treasures, and yet were unable to go ashore. Jeanne and Commerson were confined to investigating the seaweed that drifted past the boat. The ships had not been careened for months. Their underbellies trailed long ropes of weed and crusts of barnacles. The heavy drag did not help the ships beat to windward.

If you lay on your stomach and hung over the side, you might have seen clouds of tiny fish shimmering in and out of this 'under' growth. Or perhaps noticed larger shadows lurking – great triangles of fish, flat and dark with heavy beaks. Sound travels well beneath the waterline. With no high-pitched whine of outboard motors, nor thumping of passing engines, you can hear all sorts of distant oceanic messages. And in the quiet of the night, more mysterious sounds disturb your sleep at sea – the distant echo of whale song, the crackling pops of pistol prawns, and the disconcerting sound of teeth grinding against the hull as if they might eventually chew their way into the ship itself. Small wonder sailors are so prone to superstition.

Despite their stay in Tahiti – despite all the coconuts and bananas and pigs – they knew that their food would not last. Fresh food only kept so long in the interminable heat. These lands of perpetual abundance had no need of stores to see them through cold hard winters. Nature provided food, warmth and water, all year around. And what nature gave, it also took away. Within hours, fruit decayed, meat decomposed, mould grew, timber rotted and water putrefied. Nature did not tolerate stockpiles, returning every excess back to the earth. They needed to get to a European port.

They sailed through the islands we now know as Samoa, and then Vanuatu, trying to attach their observations to those of Portuguese sailor Pedro Fernández de Quirós, whose journal Bougainville was following. They made reference to the past, to stories and suppositions, to pencil marks on scrappy half-empty maps, as if it was the solid earth in front of them that was not real.

The people here were different from the Tahitians, not as handsome nor so friendly in the Frenchmen's opinions. When Ahutoru tried to talk to them, they did not understand. He dismissed them, telling Bougainville that he should just shoot them. To be fair, I suppose that was how most encounters with Europeans ended, no matter what their intentions. The Tahitians learnt this with Wallis, were reminded with Bougainville, would find out again with Cook. So too would the people of these islands. And perhaps it was how many Polynesian interactions ended too, in bloodshed.

The reefs that surrounded these islands bared their white teeth against the intruders. Fires burnt along the shores as they sailed past. Drums beat across valleys. Sailors are meant to be good at reading the signs, but they remained obtusely oblivious to these. The canoes came out, warily, laden for trade, in a tentative effort at preliminary diplomacy, but the French had nothing to offer but scraps of red cloth.

Bougainville sent some small boats ashore to search for wood and vegetables. They buried a plaque claiming possession of these islands for France. They did not ask permission. The locals objected and shots were fired. Three of the Vanuatuans died and several were wounded, although Caro was not sure how many. The visitors loaded their timber, cleaned their muskets and pondered the mistrustful and warlike nature of the residents in these parts.

The Tahitians had lit fires on their reefs to guide their canoes to safe water after dark just as the Greeks and Romans had built illuminated beacons at the entrances to their ports. The French relit these navigational innovations in the sixteenth century after the darkness of the middle ages with the Cordouan lighthouse, near the Gironde estuary at Bordeaux, over time incorporating technological advances such as parabolic mirrors, clockwork mechanisms, the Argand lamp and the Fresnel lens. Cape by cape, the coasts of the world have lit up with beacons to guide the network of ships that traverses a watery globe, maintained by dedicated men, women and families who have lived on the most remote and isolated outcrops to keep the lights on for others

My mother took celestial navigation classes when I was in primary school, jovially encouraged to 'follow the class' of all males. It made the local paper when she took out the top grade. On *Caliph* we used much the same tools as Bougainville had: paper charts, ruler, compass, dividers, pencil, sextant and a vast array of mathematical tables. Some of our old Pacific charts still carried Bougainville's name, the cartography unchanged from the contributions of his voyage.

But in the course of my lifetime navigation has been revolutionised. We sail today by feel, to the electronic pulse of a satellite transmission, pulled like marionettes across the ocean, around reefs and rocks, by invisible strings in the sky. We watch a small screen that maps our path, rather than scanning the far horizon. Massive supertankers ply interoceanic trade routes with crews of no more than a dozen, confined within steel soundproofed bunkers, tending the internal needs of the ship with little thought for the external world. Whales, debris, small craft are all mobile hazards that are not plotted on maps – not big enough to do significant damage. The supertankers plough through them without even registering. Smaller boats can purchase beacons that transmit a proximity alert, attempting to communicate with the giant tankers electronically. They should set off a warning on the ship's navigational control panel – if anyone is watching it. Once these ships would have at least been restricted to shipping lanes, but now they wander at will to maximise the winds and weather for greater efficiency.

You can plot a course from New Zealand to Fiji on a GPS just like you do in a car. You don't even have to steer. The GPS pilots the boat down the river, through the channel, across the bay, past the ferry, between the islands, through the heads and into open water. All you have to do is watch for other craft.

As you head offshore, the sea picks up, the swell rises and the boat begins its steady canter over the waves, a rolling gait that will, with any luck, remain calm and steady for the remainder of the journey.

The GPS draws a straight line over 1952 kilometres direct to Fiji. You sail across the South Fiji Basin, a thalassic plateau enclosed in a triangle of mountains and trenches shaped like a harp with Fiji at the apex. Your route travels parallel to the Kermadec and Tonga Trenches that link New Zealand to Tonga and Samoa, crossing the underwater mountains that run from New Guinea, to the Solomon Islands, to Vanuatu to Fiji. These changes of terrain go all but unnoticed except for a blip on the sonar registering the changing depths from 4 to 5 kilometres. Occasionally someone notices a mountain peak beneath them and throws a line off the back of the boat. The steel cables flicker in the sunlight, the lure spinning and flashing in the wake as the ship drops its speed to a more enticing pace. The float disappears and a metre-long Spanish mackerel (*Scomberomorus commerson*) flashes to the surface.

In the middle of the voyage you are 976 kilometres from the nearest land, with 480 kilometres of atmosphere between you and outer space, and 4 kilometres of water beneath you. It is the 4 kilometres that bothers you most. It's not really even that deep. The Kermadec Trench is 10 kilometres deep, not much shallower than the deepest known point of the ocean at the bottom of the Mariana Trench.

It's not the nothingness that bothers you – the nothing between here and the next piece of land, or the nothing between you and the Milky Way that sprawls like a glowing carpet in the pristine darkness. It's the something that might be lurking in the depths, the unknown creatures swimming beneath, an unseen, inexplicable world living just out of sight. Is it that, like Herman Melville, you give too much consideration to the way 'its most dreaded creatures glide under water, unapparent for the most part, and treacherously hidden beneath the loveliest tints of azure'? Or is it that we are mesmerised by the bow-crushing might of a white whale rising from the waters, or shark jaws festooned with rows of razor teeth, or tentacles of giant squid with unblinking eyes rising beneath a moonlit sea? Kraken, Leviathan, Cetus, Hydra, Scylla, Moby Dick, Timigalam, Yacumama – our imaginations are filled with creatures from the depths.

There is nothing to do in the middle of the night, alone at the helm, but stare at a blinking light on the screen and think dark thoughts. There is no internet signal here, no phone towers. Communication with the outside world is curtailed to a few snatched moments of priceless satellite traffic. There is no task to keep your focus, no sails to watch, no course to steer. The compass hovers magically on its setting, swinging only mildly from side to side, always returning like a butterfly on a string, more precise than any human hand or brain could ever manage. The ship steers its own course and all you have to do is stay awake for your allotted hours and keep your mind from drifting into other places.

The tip of Fiji's Tomanivi appears on the horizon, an extinct volcano, cloaking itself in cloud rather than heat-spumed ash.

As you pass the smaller southern islands of Kadavu and Vatulele, the ship shows no sign of slowing, no deviation from its 2000-kilometre route, but powers up along the coast before suddenly throttling back and changing course north-north-west, towards Nadi. With centimetre precision, the ship slides itself along the channel south of Akuilau Island and turns sharp to starboard, following the breakwater and navigating around the mudflats into the Cove at Denarau. Only then, in the midst of luxury cruisers, do you turn off the GPS and human agency retakes control to dock alongside the floating pontoons.

Quirós had piloted the voyage of the Spanish navigator Álvaro de Mendaña de Neira, who twice sailed from Peru in search of Terres Australis. His first voyage had charted the Solomon Islands and Tuvalu, and he returned with a second expedition of settlers, including his own wife and her family, visiting and naming the Marquesas Islands on the way. The settlement on Nendo Island, which they called Santa Cruz, failed. Mendaña, and then his brother-in-law, died, leaving Quirós and the new captain to continue on with the ships and survivors to the Philippines.

It is Quirós's account of the journey that we remember. He wrote and published his journal of the voyage, which was referred to by all navigators in the area thereafter. The captain who brought the remaining 100 men and women of this famous expedition back to the Philippines under stern and uncompromising command is all but forgotten. If she wrote a journal it was not published. There is no mention of Dona Isabel Barreto de Castro in the accounts of the navigators I have read – only her pilot Quirós and her husband Mendaña. This woman, who crossed the Pacific several times, who was one of the first women in history to hold the title of admiral, barely rates a mention.

Curious that the paths of these two women with such similar surnames, Jeanne Barret and Isabel Barreto, should cross in the middle of the ocean, both leaving as little trace in the written record as their ships did in the sea.

The *Boudeuse* hung mid-Pacific Ocean, square sails brailed against the spars. Only the main topsail shivered in the light breeze. There was nothing ahead but the horizon swinging in a perfect arc from north to south. On a sunny day, the water was so clear you might think you could see all the way to the bottom of the ocean, but no lead line on this ship, no anchor rope, could find the bottom here. There was no land ahead and none below either.

Dark cloudbanks formed mirage mountains that dissolved into passing squalls, stippling the flat metallic sea. The great forested peaks, which yesterday thrust themselves from the skyline like Protean sentinels, had subsided behind them. No white sandy atolls here, no canoes laden with fresh fruit and pigs. No rocks slipping just beneath the surface, nor coral reefs with teeth bared either. Just a ship in limbo rocking on its keel, waiting for its slower consort to finally catch up.

There was nothing much happening on the *Étoile* worth reporting. On 23 May, Commerson noted that the *Étoile*'s clerk, Michau, had been arrested and locked in the brig. He offers no explanation for this, nor any hint as to why this was worth commenting on. I wonder what the otherwise unremarkable Michau did to deserve his punishment.

Saint-Germain – the clerk on the *Boudeuse* – had other, more scandalous, news.

'It had been long suspected that M. de Commerson, a botanist doctor aboard the *Étoile*, had a girl for a servant, whom he had embarked at Rochefort,' wrote Saint-Germain, 'but today, it's no longer a mystery. Various young people visited him despite all the precautions, complaints from Mr. de Commerson, and he really is a girl.'

Bougainville, watching impatiently as the *Étoile* crept ever closer, chafed against the delays his slower consort caused. The islands of these waters, so rich and bounteous, offered only a transitory reprieve from their hunger. The feast of fresh food they enjoyed in Tahiti had already receded, like that of Tantalus, leaving nothing for tomorrow. They would soon be forced to fall back on three-year-old horse beans, bacon and salt beef.

Already the signs of scurvy were reappearing. Every soul on board this ship was the commander's responsibility. He was determined not to lose a single one of them.

It had been 539 days – one year, five months and 22 days – since they left France. Nearly ten months since they last saw a European face, in the Spanish colonies of South America. They all longed to make new discoveries, to make their names, to find new lands and chart the east coast of New Holland, completing the circumference that would become Australia, but they also longed for the familiar and the known. Glory must always be weighed against the risks. Whatever they achieved, they would, in any case, be condemned for not having found more, done better. Bougainville preferred to return home with all his men.

The *Étoile* was making less than 2 knots in the still air, but finally they were close enough to hail, for the dinghy to be lowered and for Bougainville to clamber aboard and make his way to his storeship, to attend to 'some business' in the captain's cabin. It had been seven weeks since they left Tahiti and the last few days of calm had, at least, given the opportunity for interacting with the *Étoile* more frequently. Each ship and crew had their own peculiar challenges and imbalances. Sometimes they could be left to sort themselves out, sometimes they could be nipped in the bud, and sometimes they lingered on and eventually had to be dealt with directly.

Was ever a captain's cabin home to a stranger collection than this one? Boxes and books, papers and presses, specimens in all stages of drying and decomposition, emanating a heady aroma of fish and sulfur, mixed with the sweet herbal scent of dried plant matter.

Who knew which of these plants contained the seeds of new food crops, new trading empires? Which of these shells formed the currency for new markets – for food, pearls, nacre or slaves? The turban shell from Tahiti might have been an overgrown version of the tiny *phasianelles* found on any French beach, except that its patterns were more precise and intricate, like a tapestry, or bands of ancient indecipherable text. The slightest chip on the surface revealed a layer of iridescent green. Everything in nature was more spectacular, more peculiar, more vibrant, here in this Pacific Ocean. It was a naturalist's heaven.

Jeanne, the cause of all the controversy, entered the cabin. As a man, she did not stand out among the hundreds of men on the ships, and were it not for the gossip that trailed her Bougainville might well have been hard pressed to pick her out from the others. Rumour was rife on any ship. One or two hundred men crammed aboard a small ship for months on end have an insatiable thirst for tittle-tattle and gossip to relieve the tedium of long dull hours at sea. La Giraudais had clearly done his best to suppress their enthusiasm with threats and penalties, but the events in Tahiti had brought things to a head and they could be ignored no longer.

Looking at Jeanne with an impartial eye, Bougainville described her as small and slightly plump, no more than 25, with a freckled face and smooth chin. He noted that her voice was a little high – she might sing countertenor – or perhaps contralto.

It seemed impossible to Bougainville that this unassuming young man could be a woman. 'Baré', as he called her, was well trained, and her knowledge of botany was impressive. She served her capricious and demanding master dutifully and without complaint. Her strength and courage were often remarked upon. She had slogged through the icy mountains in the Strait of Magellan, carrying all of Commerson's provisions, weapons and notebooks, with neither ill effect nor ill will. Commerson laughingly called her his 'beast of burden'. Bougainville had been a leader of men for long enough to know the scarcity of such stoicism. He could do with more men like Jeanne.

Even so, Bougainville could not ignore what happened in Tahiti. The Tahitians must have thought these small pale European men were a little stupid not to have noticed a woman in their midst.

Sometimes it was convenient for a captain not to know about such rumours. Up until now there had been nothing in Jeanne's behaviour to disrupt the smooth running of the ship. But it was forbidden under naval ordinances to bring a woman on board ship. If Commerson had broken the law, then Bougainville must know and punish him.

He asked the valet directly, for the truth.

Tearfully, Jeanne confessed to everything. It was all her fault, she claimed. Commerson knew nothing. She had deceived him by appearing in men's clothing at Rochefort before boarding. She had worked as a valet before, for a Genevan. She was an orphan from Burgundy and, when the loss of a lawsuit had reduced her to penury, she decided to disguise her sex. She knew, when embarking, that it was a question of circumnavigating the world. The notion of such a voyage had excited her curiosity.

If Bougainville did not believe everything she told him, there was no reason for him to delve further. If he doubted

Commerson's innocence in this charade, he did not need to question it. No great crime had been committed. There were no naval ordinances about women dressing and signing on as men. It was not a problem that commonly arose – at least as far as anyone knew. Jeanne had been an exemplary, hardworking member of his crew. Bougainville clearly admired her determination and the fact that that she had always behaved with the most scrupulous correctness. The same could not be said for everyone aboard the ship. But since the rumours of her 'discovery' it had become increasingly difficult to prevent the sailors from sometimes 'alarming her decency', as Bougainville put it. He would have to take steps to ensure that she suffered no further unpleasantness.

Bougainville added nothing further to his account of this remarkable woman. Others on the ship were astonished by her bravery and thought she deserved to go into the annals of the world's most famous women. But Bougainville did not mention her again, beyond a brief entry in his journal noting that when Commerson disembarked at Mauritius he did so along with 'his valet, a girl as a man'.

The first woman known to have circumnavigated the world warrants barely more than a page and a half in Bougainville's journal. This is the only record we have of her own words, transcribed by her commander. But for this brief mention, Jeanne Barret, and her achievements, would have vanished entirely from the historical record.

12

IN THE TEETH OF THE REEF

The Coral Sea, June 1768

None of them knew what lay ahead as they sailed due west along the 15th parallel south of the equator into increasingly uncharted waters, least of all Jeanne. Did her revelation as a woman change the way she looked and behaved? Did it change the way others looked at her, behaved towards her? Perhaps she was the same person she had been all along. Perhaps she had never been in disguise at all. Maybe she had always been true to herself, dressing as she wished, behaving as she wished, getting on with the work and life she wanted to have. Who is to say any of it was a pretence, so much as a revelation of truth. Maybe it was being forced into dresses, confined to the kitchen and the bedroom, treated as property for trade, purchase or exploitation, that had been the falsehood. Perhaps it was Ahutoru who had guessed at her true nature most accurately – one who fell between the strict binary definitions of her times.

How her compatriots changed in their attitude is less certain, for all Bougainville had ordered that she was not to 'suffer any unpleasantness'. How strong would his protections have been once the memories of Tahiti had faded?

Vivez claimed that now that her true identity was known, Jeanne was finally able to reduce the amount of linen or rags she used to bind her breasts, making life more comfortable for her. I suspect he was giving more thought to her breasts than she was. I can see no reason why she would have felt more comfortable revealing her figure now than she would have before. She had no women's clothes to change into and there was nowhere to obtain any. For the moment she remained as she was, dressed as a man and behaving like one of the men. There was no reason nor opportunity to change until they returned to some form of civilisation.

The land that lay ahead offered no recognisable prospect of that. They all knew that this land was a ship's graveyard. The stories of the *Eendracht* and the *Batavia* on the west coast were etched into their collective consciousness. No-one wanted to be the first to mark the east coast with their skeletons.

Near midnight on 4 June, bright moonlight highlighted breakers ahead. The topman spotted a low sandbank. The *Boudeuse*, always in the lead, changed tack and signalled danger to the *Étoile*. By 8 am it was light enough for the two ships to approach more closely, within a league and a half. The low sandy island barely rose above the water, invisible except from the very top of the mast. It was covered from one end to the other with birds. They rose in a great cloud – tropic- and man-o-war birds as well as gannets, boobies and petrels.

Bougainville named the island Diane Bank. Its sudden appearance in an empty sea was ominous and put everyone on edge.

By afternoon, someone thought they had sighted land and breakers to the west. But they found nothing as they continued, laying to overnight before resuming under full sail the next day. It was a full 24 hours before telltale pieces of wood and fruit drifted past. The swell had fallen significantly and the wind had changed direction to a fresh south-easterly – all unmistakable signs that land was near. By the afternoon of 6 June, another sandbank appeared and the *Boudeuse* signalled a change of course from due west to a more cautious stepwise progress, north then west, north then west. This advance did not last long. The lookouts saw breakers not more than a league and a half away. The pounding foam stretched north to south, on a slight easterly slant, as far as the eye could see. Plumes of spray flew violently into the air, and here and there the jagged teeth of rocks or coral appeared beneath the bared lips of the waves. A distant thunder drifted on the wind towards them, even at this great distance. No-one wished to go any nearer to the seething Charybdis.

It was a sign from God, someone said. They could proceed no further on their westerly course at this latitude. Beyond the great lathering shoals lay 'New Holland', of that there was no doubt.

I draw a line around a globe on the 15th parallel from Peru across the Pacific Ocean. There is no land until you reach the islands and atolls of French Polynesia, Samoa and the islands of Vanuatu, and then nothing until you reach the Queensland coast, just south of Lizard Island. Keep going across the Australian continent and you slice through the Gulf of Carpentaria then across to Western Australia in the Kimberleys. From here across the Indian Ocean there is no land until you reach Madagascar and then Africa. And in the Atlantic there is only the island of Saint Helena, sliding past, slightly to the south. It reminds me of just how small the chances were of finding any land in crossing these oceans.

If Bougainville's ships had kept approaching the coast, and had found a way through the reefs that lay between them, they would have arrived just south of modern-day Cooktown. At its most southern extremity the outer reef is 250 kilometres offshore, providing a wide clear passage of calm waters, beloved of sailors and tourists across generations. But as you go north, the Great Barrier Reef grows progressively closer inshore. Off Cooktown, the outer reef is only 40 kilometres from the beach. The inshore passage, from the Daintree northward, is a treacherous labyrinth of coral. It was no place for a sailing ship.

Bougainville was 140 kilometres out to sea off Cooktown. In three years' time, Cook would come to grief here in the *Endeavour*, on the inside of this same tangled maze. Bougainville was locked out, Cook was trapped within. Cook finally escaped, squeezing out through a narrow passage between 'Half Mile Opening' and 'One Mile Opening'. It is the last point before the reef closes in towards the Howick Island group close inshore. Even today, in shallow-drafted boats with motors, the passage north is tricky and dangerous, requiring careful navigation and constant vigilance.

The *Endeavour*, the *Boudeuse* and the *Étoile* are ghost ships passing in time, following each other's faded wakes across distant seas, each seeking to correct, refute or supersede the discoveries of those who came before, or forestall the claims of those who come after them.

The French ships were already further west than Quirós and on the same latitude as Dampier was when he abandoned his exploration of the west coast for lack of fresh water. They had no reason to think they would find any more hospitable land approaching from the east. Dampier himself had suggested that Australia (or New Holland, as he knew it) was no grand continent at all, but rather a cluster of islands, and this hostile sea full of shoals and sandbanks gave them no reason to think any different. There was likely nothing to be gained by making the coast, and great peril in doing so.

The *Boudeuse* signalled a change of course to the north-north-east, heading to Java.

I had always thought that Bougainville came within sight of the Australian mainland, that he had described it as lush and verdant. I am certain that I read that somewhere, have perhaps even repeated it, but I can't remember where and it was clearly a mistake. The coast is indeed lush and verdant here but they were too far offshore to see it. The Australian coast is low and flat. The Great Dividing Range peters out in the northern reaches at 1611 metres, on Mount Bartle Frere south of Cairns. I lived here for a few years, in Innisfail, in the shadow of this mountain, when we stopped for me to finish my last two years of high school at the local school. We bought an old Queenslander on the river and moored *Caliph* in the muddy mangroves against the bank. I went to school, wore a uniform, chafed at the regulations and wasted time, made friends and took up rowing. From our wide verandah we could sit and watch the lightning crack across burning canefields as we waited for the rains to break. Mount Bartle Frere was visible in the distance, at least on a clear day and, though distinctive, it is not particularly imposing. I don't recall, on any of our trips to the outer reef, that it could be seen very far off the coast.

It is easy to mistake the ship's location in Bougainville's narrative, with unclear bearings and no demarcation of coasts or landmasses. Maps of sufficient size and scale to reveal the precise details of location are expensive and unwieldy, needing to be folded in vast sheets and tucked into special pockets in the back of hardback books. They are the first luxury that publishers dispense with, leaving us to rely entirely on the imperfect imagination of a written description. I am delighted when I open the pages of French historian Étienne Taillemite's maps reproducing Bougainville's voyage. The sheets stretch across my wall, his route precisely marked in purple dashes and dates.

A map cannot be replaced by words. French philosopher Bruno Latour has argued that it is not the map itself that is important – you can, after all, keep a map in your head, redraw it at will. It is the mobility of the map, the capacity to take it elsewhere and have someone know the shape and geography of a country they have never seen, to answer questions and 'determine who was right and wrong'. Not just over space, but also over time.

Bougainville did not fill in the gaps on his maps with imaginary lines. He left the spaces blank, with tiny sections of coast along his path, dotted through an otherwise empty Coral Sea. He could sense the east coast, the presence of a significant landmass, through the difference in the swell and the winds, the changes in wildlife, the passing of debris – wood, wrack and banana leaves – but he did not map it and did not pretend to do so. He laid no claim to this coast.

I have sailed the route that Bougainville took north across the Coral Sea. I had been away at university but came home to help my dad sail *Caliph* to Papua New Guinea, where he'd been chartered to carry cargo between the smaller islands. We started further south, taking the Grafton Passage out of Cairns and a direct path, almost due north, towards Samarai on the easternmost tip of New Guinea. Our route took us directly past Bougainville Reef. Our invisible paths intersected, 221 years apart, at about the place Bougainville decided to head north.

Our preparation for the journey had not been ideal. We had spent what felt like a small fortune at the supermarket, stocking up on giant tins of beans and soup, dried milk powder and jam, fishing hooks and sweets for trading. We had a refrigerated cool box mounted on the back deck, the first time the boat had ever had refrigeration instead of only an icebox. Feeling decadent we stocked up on fresh milk and cheese, eggs and meat – enough for the first week of the journey at least. Seduced by the overcast conditions, I forgot to protect my pasty southern skin and burnt my back and shoulders raw and blistered. A day or two before we left, an electrical circuit in the engine room overheated, burning out all the wiring to the batteries and filling the hull with the choking bitter smoke of charred plastic.

Despite these setbacks, we departed on time, and carefully picked our way out of the reefs with the aid of detailed charts and clear, well-placed navigational beacons. By nightfall, we were in open waters, settling into the regularity of two hours on, six hours off between our four crew, darkness and a sea breeze offering some welcome respite from tropical heat.

The breeze, however, was not enough for sail alone, so the engine thundered constantly below deck, maintaining a suffocating heat amid the diesel fumes. During the day, the sun seemed to hang high overhead for far longer than its designated midday hours, shearing any shade from the sails. The momentary cool of the ocean evaporated in seconds, leaving a crust of sticky salt across burnt flesh. I slathered my back and shoulders with aloe vera, soaking the peeling layers of skin into a putrid mess that smelt like the flesh was rotting from within.

On the third day, the refrigerator died. We gorged ourselves on consumables and stored what we could under the floorboards in the bilge where the temperature sat in the mid-twenties. A huge block of cheese, wrapped in a tea towel, lasted the longest, progressively getting softer, gooier and smellier as it matured.

I reminded myself of my French ancestry, blocked my nose and ate it anyway, treading the fine line between gourmet and gross.

No-one wanted to spend much time below deck preparing food, so meals reduced down the minimal efforts. A New Guinean favourite, from a country not renowned for its cuisine, became a staple: rice fried with flaked corn beef and a little curry powder. Thirty years later I still crave this irreproducible dish, but at the time I found myself dreaming, like Rose de Freycinet, of elaborate dishes I could not have.

Glittering days passed with gritted teeth and squinting eyes – broken only by a crab swimming on the surface of the open sea, or a loggerhead turtle raising its head as we passed. A sea snake swam past in the opposite direction, writhing black and yellow stripes, apparently oblivious to our presence. They are highly venomous for all they lack the hypodermic syringe action of the more familiar terrestrial Elapidae snakes. In Australian terms they are only moderately worrying. Another swam past, then another and another. I counted 147 before they petered out. I have no idea where they were going or why.

I sat on the chains under the bowsprit, my feet skimming the water. I hoped for bottlenose dolphins to join us, as they often do in southern waters, but these northern species seemed less sociable and we saw them only in the distance, skittering small and dark across the surface. I spotted something white floating on the water in the distance and signalled the helmsman to bring us closer. It could have been rubbish, a broken piece of polystyrene, a worn and sunbleached buoy from a fishing net or mooring. But there had been very little rubbish here, in this cul-de-sac between Australia and New Guinea, away from the great oceanic gyres that pool islands of garbage into their hearts.

As we came closer, I climbed down into the chains of the dolphin striker and scooped the object out of the water as it bobbed past the hull. Remarkably, it was a chambered nautilus, empty of the squid-like animal that once regulated its buoyancy within the water column. Rare and hard to come by, nautilus are mysterious creatures of the mid-ocean, usually found below 100 metres. Nowadays you can call up video footage of them online and watch them clunking awkwardly against one another while they scavenge on decomposing flesh. Their unblinking eyes stare out from an exquisitely patterned mantle, their tentacles neat and tidy as a trimmed beard.

This nautilus was one of the last survivors of the great dynasties of nautiloids, ammonoids and coleoids, nearly all of which were wiped out by the Cretaceous mass extinction along with the dinosaurs 66 million years ago. Now all we have of these once dominant species are their shells, layered in abundant fossil beds across ocean floors, shaped in coils, spirals, cones, hooks and even paperclips. A friend gave me a fossil ammonite from Nepal, its imprint trapped in a smooth black shale that unlocked to reveal its two-sided treasure, connecting me across 100 million years to this pristine undamaged specimen floating in the ocean. The friend had told me at the time that they were also found opalised in the Australian desert, which once formed a shallow inland sea, a notion so intriguing and puzzling that it would one day lead me to write a book about it. Not for the first time, I wonder how many of my books are driven by the questions I could not answer in my childhood.

In the distance, the mountains rise above the horizon, steadily growing as the hours tick past. We comment on their amazing height, wondering how long it will be before we can see their feet dipping in the ocean. The newly minted highlands of New Guinea, a fold in the Australian continental plate thrust over 3 kilometres skyward by the continent's relentless slide north over the Pacific plate, are mountains on an entirely different scale to Australian eyes.

'Wait, look, up there,' one of the crew says, pointing. 'Above the clouds.'

We look up above the thick layer of clouds encasing the mountains and realise their peaks emerge sharp and dark, impossibly, improbably lofty, from above the clouds themselves. Higher than the Alps, only here they seem to rise directly from the sea itself. I could not even have dreamt, could not have imagined, that such mountains were even possible.

Bougainville described this country as one of the finest he had ever seen, with mountains so high they were lost in cloud. Did these darkly wooded slopes remind Jeanne of the smaller mountains of her home? Or did they seem implausibly alien in all their vast stature to her too?

I am increasingly frustrated by Commerson's journals. I had expected accounts of the species he has found, of plants collected and sights seen. But even the naval officers make more comments on nature than Commerson does, even it was just 'Saw lots of birds.'

I know that Commerson was saving his biological work for publication, that many naturalists kept their science separate from their travel narratives. I am not quite sure why. Natural history narratives were popular in the eighteenth century – Georges-Louis Buffon's *Natural History* from 1749 is often said to have outsold Voltaire, making him one of the most widely read authors of the century. But travel narratives did not seem to include nature until the nineteenth century. I long for Alfred Wallace's vivid descriptions of collecting under seemingly impossible conditions in the Malay Archipelago, battling with dogs and ants for his precious specimens.

Or Darwin's thoughtful considerations on the origins of life from the *Beagle*. Both of them mention their assistants and helpers regularly. But I can find so little material in Commerson's writing that it's hard to even remember he was a biologist, let alone one of the greatest naturalists of his age. And of course, there is almost no trace of Jeanne. You'd think he was working entirely alone.

Like Bougainville, we missed our mark as we approached the New Guinea coast. We should have reswung the compass before we left, adjusting for changes in the earth's magnetic field. As it happened, we arrived very close to where the *Étoile* and *Boudeuse* had approached, although we were fortunate to have vastly better weather.

The coast stretched in each direction in a long low line of breakers against a rocky coast. Vast quantities of debris washed around us from the rivers that surged down the slopes of titanic mountains. We were not sure exactly where we were, or how far it was to Samarai. We could not go ashore to ask, as we had to clear customs first. Fortunately, others noticed our presence. Just as they have always done, canoes launched from the shore, now with blue tarpaulins for sails, and made their way out to us, laden with fish, fruit and vegetables for sale. We purchased corn cobs from an ancestral multicoloured breed of corn that I only knew existed from genetics practicals. We were told that we were in Amazon Bay, some distance west of where we needed to be. We were given directions and local advice, and headed east along the coast, careful to stay well offshore as night fell.

The Australian coast is studded with jewels of light, flashing their unique and characteristic signals out to sea. Count the unique beats of the lighthouses, one-two-pause-one-two-pause, and you know which lighthouse you are near. You can take a bearing, often triangulating between two lighthouses along a stretch of coast, and pinpoint exactly where you are. We scoured the horizon for the next lighthouse, staring at the dark line between sea and sky, hoping to see a tiny flash. Often it appears in the periphery of your vision, so small and so precise it cannot be seen if you look directly at it – obliterated by a tiny blind spot in the middle of your vision. So we looked to one side as if, like in quantum mechanics, the act of observing disrupted the phenomenon itself, and counted the flashes we could not see directly.

The New Guinea coastline was dark, unpunctuated by electric lights of any kind. Here and there we saw a fire glowing on a beach, but these too died down as night fell. The steady pounding of the engine pushed us forward through the clear waters irrespective of the light winds and drifting currents. The night was clear and we cooked fresh fish over coals on the back deck as we continued on our way, sailing by the board, trusting that our charts and course were correct.

The *Étoile* and *Boudeuse* wallowed on rising swells, drifting slowly north perilously close to the lines of breakers. As day broke on 10 June 1768, someone sighted land and those on the deck could smell the pungent damp earth that drifted from a large bay opening out to the south.

It seemed like a vision. No-one could recall ever having seen a finer aspect that this. Gentle plains and groves filled low ground to the seashore. Beyond rose an amphitheatre of mountains, their summits wreathed in clouds. This was no land of dry sandy soil, destitute of water like the west coast of Australia. This land was rich and fertile, well worth exploring, if only they could find a safe passage through the reef. But they did not have the supplies to make such an expedition. They could not guarantee that they would find food or water ashore. These waters were not safe for them to stay.

And yet the wind still forsook them. The swells swept them slowly sideways, closer and closer to the shore. By four o'clock in the afternoon the ships lay just off the coast of a small island, at five the commander ordered the ships to bring their heads up. Every inch of cotton was hoist aloft in an effort to catch the slightest breeze. The heavy main sails sagged, bereft of air, while the topsails puffed with intermittent enthusiasm but little effect.

By the next day, they were just four leagues off the coast. The island's residents sailed along the coast watching the distant spectacle, while behind them great fires burnt along the coast. One of the crew caught a shark and, cutting it open, found a turtle inside.

The weather grew worse. The wind picked up, blowing against them, rain fell and a fog descended so thick the ships had trouble keeping touch with each other. The sea rose in anger. The two ships had no choice but to drop their sails to avoid being driven ashore, and yet this made it even harder for them to tack and weave their way off the coast. The *Boudeuse*, with finer lines, had a better chance of beating to windward, but the *Étoile*'s bluff bows made it impossible to point into the wind. They tacked back and forth, barely even maintaining their position.

The shallow seas swept sand and weed onto the foredeck. Bougainville stopped taking soundings. He set their course as close to the wind as he could and the ship pounded 'by the board' out to sea – following a straight line on the map, oblivious to all dangers. They closed their eyes and prayed. There was no point seeing what lay ahead: death or salvation.

'We are sailing here like blind men,' said Caro. 'We do not know where we are . . . I would be a fool if I ever have to go around the world a second time.'

My paper chart comes from my father. It is crisscrossed with other voyages. Charts are expensive to buy and these ones have been in heavy rotation, borrowed and returned by successive travellers seeking to sail up the Queensland coast. Most stop at Cairns, after which the waters become precarious. Supplies and facilities are hard to come by. Beyond Cooktown there are few places to stop for food and water, and such supplies as exist are horrendously expensive.

Bougainville had just the same concerns, although his supplies had to last much longer. They had bread for two months and the salt meat was rancid. They could not get close to the *Étoile* to resupply *Boudeuse*. What poor food they had was under constant attack from rats, weevils and cockroaches.

Fifty years later, the naturalist René Lesson on the Louis Isidore Duperrey expedition described a plague of cockroaches on the *Coquille*.

'The latter disgusting insects had multiplied so much through the ship that they turned our narrow cabins into torture chambers,' he said. 'Spreading in their thousands, they fouled the food, soiled the water and disturbed our sleep. They attacked everything, not overlooking the inkwells, which they emptied without leaving a single drop, and even leather succumbed to their appetite . . . Mr. d'Urville and I gave up our cabins and for more than ten months, for as long as our voyage lasted, we slept on a native mat stretched out on the deck from which even the rain could not drive us away.'

We counted ourselves fortunate not to have cockroaches on *Caliph*, but in the tropics I occasionally slept on deck, even in the rain, when nights were so hot and still that the alternative was to drown in a pool of sweat below deck anyway.

I remember, too, the careful stock book on *Caliph* recording the supplies packed tightly into sealed containers stored beneath the floorboards. We crossed off each tin of beans or tuna and monitored the slow decline of rice and flour. A friend once told me of an ocean crossing with her parents that took longer than expected. She could see their stocks were getting low and with a child's logic she thought it better to stop crossing off the items on the inventory than admit how little food was left. In the end, they survived the last two weeks on brown rice and tomato sauce.

The food situation on the *Boudeuse* and the *Étoile* was far worse. On the *Boudeuse*, they ate the goat that had been on board since the Falklands and had given them a little milk every day. A young dog that Nassau-Siegen had bought in the Magellan Straits was sacrificed. Someone even killed the ship's cat but superstition prevented them eating it.

'He was killed for food,' said Saint-Germain. 'Some sailors claimed that they had seen it fall from the top of the masts, so it was thrown overboard.'

I wonder if their deteriorating health made things better or worse for Jeanne. Whether starving men were more, or less, dangerous than well-fed ones.

They ate their last pig on 3 June and started eating the untanned leather the grain had been provided in. They even started eating the leather in the rigging until Bougainville banned the practice for fear the ship would fall apart. They soaked it and scrubbed it of its bristles but it still didn't taste as good as the rats. Roasted ship's rat turned out to be quite a delicacy.

'We found it quite excellent,' wrote Saint-Germain, 'and we shall be happy if we can find some more before others acquire a taste for them.' Later he ate three rats for dinner 'with great gusto'.

The situation was not much better on the *Étoile*, even though they had more supplies.

'We had to reduce [rations] to twelve ounces of bread, two ounces of vegetables and three quarts of water, as our sick were increasing considerably,' wrote Vivez, 'their number including twenty scurvy cases with nothing to give them but water and bad rice and even this in small quantities. The frightening uncertainty about our ability to leave this abominable place where we have no hope other than the tidal currents that seem to favour us.'

Vivez did not bother to describe the islands and reefs they were sailing through, north through the Solomons to New Ireland and New Britain. To his mind there was no point.

'No-one would want to come here,' he wrote.

13

NATURE'S BEAUTY AND TERROR

Port Praslin, July 1768

WHEN WE SAILED ALONG the north coast of New Guinea in the late 1980s, few of the villages had roads. Their traffic was mostly by sea, in small boats and canoes or by foot along narrow muddy tracks, barely big enough for two people, that trailed up over headlands between the beaches. At high tide the water rose beneath the thick overhanging vegetation and foot traffic splashed through knee-deep waters along the invisible beach, or waited for the tide to turn. The isolation kept the villagers safe from unruly 'rascals' in the towns, although it was never possible to arrive in these villages unexpected. Our presence was telegraphed along the coast from boat to canoe, from village to village.

Canoes floated on offshore reefs: heron-like sentinels standing in watchful anticipation of fish and noting the approach of unfamiliar vessels. Those with fish to sell sped towards us, while others paddled back to shore or slipped swiftly away under blue tarpaulin sails.

The mirrored waters reflected the forests and mountains above so perfectly that it was sometimes hard to be sure where land ended and sea began. Territory here, despite the French origins of the word, slips along the edge of the coast and extends far out to sea. There was no mistaking when we crossed into each of these aquatic territories, around a headland, into a bay. Unlike in Australia, the seas here are not some kind of amorphous national or international waters giving free passage to any who sail upon them. Here the water is owned, claimed, possessed and defended. As we approached we prepared for the inevitable inspections, the expressions of mutual respect, the dutiful exchange of goods. It is an obvious and unavoidable courtesy.

And yet the *Boudeuse* and *Étoile* ploughed on regardless across ancient tribal boundaries, swatting away local resistance as if it meant no more than the irritation of mosquitoes that plagued them whenever they came too close to shore.

The rain fell relentlessly, day and night, on their arrival at Port Praslin, not unlike standing beneath some celestial waterfall never-ending in its generosity. Even when the weather seemed to pause to draw breath, it reconsidered, sending sudden dark squalls scudding back across the bay.

For all the inclemency though, this bay on the southern tip of New Ireland was a safe one: a secure anchorage surrounded on three sides by steep heavily wooded slopes, an abundance of freshwater running to the shore and – best of all – it seemed uninhabited. The people of these islands had made no secret of their antipathy towards visitors. Sometimes they arrived fully armed in their canoes, paddling at great speed with blood-curdling cries – bows raised, ready to strike with piercing arrows. Other times the islanders waved them closer with branches of peace. They grinned as they approached the ships, teeth stained red with betel juice, before suddenly striking with spear and arrow. No warning shots deterred them – only bloodshed caused them to cry out in surprise and retreat. They hovered, just beyond musket range, shouting and rattling their spears, flashing their bare arses in a universal gesture of contempt.

But here, there was no-one. Those on board the ships felt safe from other people, if not from each other. There was no escaping their own kind, not even on the far side of the world. They found the remains of an English camp at the end of the bay – just a few months old – from the same ship that had preceded them in the Magellan Straits, in Tahiti and now here. What strange magnetism drew them together, across thousands of miles in which they might have gone in any direction?

If they sometimes thought they heard the shouts of men from high up in the hills, they were left unmolested. No-one came for the small canoe they found stored at the far end of the bay, nor the camp with remains of recent meals of shellfish and boar.

Perhaps the place was deserted because there was no food. They had hoped for coconuts, bananas or other fruit that might alleviate their scurvy. They made a camp on the shore where the sick might rest and recuperate but it did them little good. What fruit they found had been picked over by pigs. The cabbage palms were fiercely defended by a swarm of giant ants. Even the fishing was poor.

'Our crews are worn out,' wrote Bougainville. 'This land is providing them with nothing more than insalubrious air and extra labour.'

They did such repairs as they could to the ships. The timber, at least, was good – some for burning and some for the masts and spars. And there was an abundance of water. No fewer than three streams plunged down the hillside onto the beach. The first was allocated to resupply the drinking water of the *Boudeuse*, the second for the *Étoile* and the third for washing.

When the weather cleared, briefly, it did so without constraint, to the fairest day imaginable. They lay on fine white sand lapped by turquoise waves and admired the waterfall from which the *Étoile* drew water, which was more spectacular than anyone could possibly imagine.

'Art would struggle in vain to produce at Versailles or Brunoy what nature has cast here in an uninhabited spot,' Bougainville declared. 'What hand would dare to build the leaping platforms held back by invisible links whose graduations, almost regular, promote and vary the spilling flood, which artist would have dared to create these storeyed masses that make up basins to receive those sheets of crystal water, coloured by enormous trees some of which rise up from the basins themselves? It is quite enough that there are privileged men whose bold brush can trace for us the picture of the inimitable beauties. This waterfall would deserve the greatest painter.'

Nature overflowed in abundance, if not utility, and Jeanne and Commerson lost no time in making their collections. The forests were filled with an infinity of insects, beautiful and bizarre. Phasmids stalked and swayed on branches – some like spiked thorny devils, others like delicate leaves. The trees were filled with parrots and beautiful turtledoves.

Bougainville, though, was more interested in the shells on the shoreline.

'The reefs that line the coasts contain treasures of conchology,' he wrote. 'I am persuaded that in this respect we have collected here a great number of new species.'

And yet he mentions only two species – an elegant heart-shaped bivalve and an abundance of hammer oysters.

I'm disappointed that he doesn't mention any other shells, and even those he does are not particularly exciting. A few years later, another French naturalist, René Lesson, provided details of the nautiloids and giant clams that are so abundant in these waters. He lists shells whose French names I don't recognise, but the oysters, limpets, nerites, conches and strombs are familiar. Jeanne must have collected shells here too. Like Tantalus reaching for waters that always recede from his grasp, I search for a reflection of Jeanne in these accounts, but all I can see are the strange wandering eyes of the stromb shells on the reefs looking back at me with mutual incomprehension.

Lesson also described huge baler shells that New Guineans used to empty their canoes of water. I have a soft spot for these big volutes, which can grow up to 45 centimetres long, with their raised siphon and their vast foot spread like an ornate picnic blanket. I used to try to preserve the smaller volutes intact when I was given them by fishermen, but their colourful feet always retracted and faded in death, no matter how I pickled them. My baler shell, the largest in my collection, stands empty: its apricot-pink interior a pale reminder of the animal that once inhabited it.

Commerson, as usual, remained silent on the matter of shells, even when everyone else was collecting them – including Jeanne. And I am beginning to wonder if I will ever find them at all.

Commerson made hardly any entries in Duclos-Guyot's diary during their stay in Port Praslin. One would think he must have been busy collecting specimens here, or at least examining the specimens Jeanne brought him. And indeed there are many drawings of birds and plants and insects labelled 'Nouvelle Bretagne' in the archives of the Muséum National d'Histoire Naturelle in Paris. The spiky phasmid described in Bougainville's journal made it back to Mauritius at least. There is a picture of it among the plates of Commerson's insects. Nearby is a tiny little spiny orb-weaver – a family of spiders so brightly coloured and bizarrely shaped that they look more like alien spacecraft than arachnids. The description tells me that it is red, so I assume it is *Gasteracantha aciculata*. A zebra moray eel pickled in alcohol, with its bold black and white stripes, is the only other specimen I can find from this region that has made its way back to France.

I am surprised they collected so few plant specimens from Port Praslin, but many of those that they did collect are now type specimens – the first of their kind to be recorded. Strange and bizarre tongue ferns and spikemosses, asplenium, aspidium and polypodium ferns, lilies, exquisite rock violets and beauty berries, trees with sea trumpet flowers, terminal buds and medicinal fruits. The ferns are particularly beautiful. They could have been pressed yesterday. Another common plant, now known as *Bikkia commerconiana*, has such long narrow flowers that I wonder what pollinating creature, moth or bird, might have a snout fine and protracted enough to reach its nectar.

One of the beautiful rock violets from Commerson's collection has found its way to a herbarium in Geneva. It is now known as *Boea magellanica*, but it has previously been known as *Boea commersonii* and also gone by the name of Beau. A note on the specimen explains why.

> The name of Miss Beau, niece of a chaplain of St Honore, who followed Commerson in the voyages, and was made known as woman by the Tahitians in spite of her disguise.

This is clearly an oblique reference to Jeanne, but somehow conflated with Commerson and his brother-in-law, Father Beau. It is an old note, perhaps from the eighteenth or early nineteenth century. The handwriting of this note could be Commerson's but I don't think it is. His name is written very differently from the way Commerson writes it, with a small capital *C* and a short *s*. And several other letters are distinctly different too. Whoever it comes from, whatever its origin, this version of the story is interesting, not just because it stresses the virtue of Commerson's assistant, but also because it links Jeanne back to Toulon-sur-Arroux and to the Beau family, a link which was only discovered in the 1980s by Henriette Dussourd.

It's another strange clue, simultaneously misleading and revealing, that leaves me even more puzzled than I was before. I wonder what other overlooked clues there might be on other specimens, in other journals. There are boxes and boxes of Commerson archives in the Muséum National d'Histoire Naturelle and elsewhere: journals, manuscripts, images, letters, scientific notes and lists, as well as thousands of specimens with tiny notes attached in herbaria around the world. I cannot read them all. Many of them are in Paris and I can only afford a couple of weeks to visit, a rare luxury. I don't have time to go through all of the archives hunting for hints that might shed light on Jeanne. It would be an insurmountable task to try to classify, catalogue and analyse them all.

I worry that I rely too much on the published records. And often these transcriptions skip material from the original journals that they think is not interesting: too technical, too nautical or too scientific. These are often the very things I am most interested in.

Sometimes even an absence of information is useful – when someone stays silent on a topic that everyone else speaks of, when someone mentions something out of turn. It is the little things – the notes that sound just slightly out of tune – that reveal the most.

On 18 July, Duclos-Guyot reported, in a few short lines, on the usual daily activities of the ships, much as he had on previous days, and much as he would do on the days after. Nothing was happening. It had been seven weeks since Jeanne's confession to Bougainville and another eight weeks since her 'discovery' on Tahiti. But at some stage Duclos-Guyot added a note in the margin.

'It has been discovered that the servant of M. Commerson, the doctor, is a girl who has been dressed as a boy.'

His pen trailed off on the final letter with a wavering and uncertain scroll, as if he didn't quite know how to finish his thoughts on this matter or if he should even have mentioned it. I am not quite sure what to make of this either.

We had barely dropped anchor off the New Guinea mainland in Milne Bay before the main fleet of canoes was launched, overflowing with a few men, but mostly women and children. They poured over our boat, peering through portholes and down hatches.

'They are surprised to see a national on a boat like this,' said one of my crewmates, a New Guinea woman who lived in Cairns and who sailed to Madang with us. 'They think only white people sail on these boats.'

She was probably right – but the fact that the crew were all women, apart from my dad, might also have been surprising. I didn't think much of it at the time – it was nothing out of the ordinary. And I was too busy keeping out of sight. My fair, straight hair was clearly strange and exotic enough for all the young girls to want to stroke it adoringly.

The young men were not so adoring. The look in their eyes

was of a different nature. Sometimes they would come aboard, helpful and deferential to my father, offering to guide us from one anchorage to another. They arrived on one occasion just as I had started winding up the anchor, a task I had been doing since childhood. There is a knack to shifting a weight efficiently and steadily, whether it is hauling on the sheets to pull up a sail or winding up an anchor – a rhythm to the heave and haul, allowing momentum to ease the load, working with the rise and fall of the ship, or using resistance to your advantage. Being small, I had learnt these techniques young, to maximise what strength I had. But these young men impatiently waved me aside, anxious to show a girl how it should be done. They yanked on the winch handle, which immediately choked and jolted with an iron bite. They glared at me as they struggled to turn the handle, as if I had tricked them by making it look too easy. The awkward impasse was only alleviated when Dad took over the task after coming to find out what had caused the delay.

It seemed strange that everywhere we went in New Guinea, there were people. It felt more crowded here than in any busy city. There was no escape from scrutiny – every step we took was watched, noted and policed. There was no section of coast, no coconut tree or beach, that was not owned by someone, or claimed by a community. I had grown up in a world where the coastline could not be privately owned, where access to the intertidal zone was almost never obstructed. On a boat, we were free to come and go as we pleased. We could moor wherever we liked, in front of million-dollar mansions, or in bays where no-one lived and no roads reached. We could step freely onto any shore and explore the coastline. We caught fish from the waters, ate oysters from the rocks, collected shells and treasures, and cooked our meals on fires made of driftwood on the beach. We came and went and asked no-one's permission.

But we could not do this in New Guinea. Here we were foreigners, white and (mostly) women. This was not our place.

I had always thought my freedoms were a right, and now I wonder how much of this is simply privilege and entitlement. But either way, I do not want to give them up, any more than did those first Europeans who sailed unwelcomed along these uncertain, uncharted shores.

It must have felt like nature, in all its vigour and vehemence, conspired against Bougainville's small fleet. The constant bad weather prevented work. The astronomer Pierre-Antoine Véron was afraid that an impending eclipse, visible only in the southern hemisphere, would be obscured. Over the course of the voyage he had made thousands of readings and measurements that greatly corrected the errors of previous navigators and cartographers. This was the first voyage on which longitude was consistently measured astronomically. On the day of the eclipse the showers cleared enough for his observations to be made. Véron used his records of lunar distance and the calculations of longitude taken at Port Praslin and the Strait of Magellan to calculate, for the first time, the vast size of the Pacific Ocean.

In late July, a sailor was bitten by a small black sea snake with a white head. Ahutoru was certain the boy would die. The young lad complained of pain, which worsened as the surgeons forced him to walk, thinking this would help him sweat out the toxins. The walking might well have saved him from further fatal paralysis. The boy collapsed and went into convulsions and they administered opium. But the next day, he started to recover, much to Ahutoru's amazement.

And then there was the earthquake. An hour before midday a rumble rose through the ships at anchor and continued for two minutes. The waters rose suddenly around the feet of those collecting shells on the reef up to their chests and then fell, as if nothing had happened.

Even the earth beneath their feet did not want them to stay.

I cannot tell you one way or the other what happened at Port Praslin on 22 July 1768. I know what I think, but others will not agree. I can only show you what was said, what was not and what was crossed out or erased. You will have to make up your own mind.

Bougainville mentioned the weather and the earthquake in his journal, even though the earthquake was not sufficiently interesting to include in the official publication.

> These 24 hours have been 24 hours of rain and bad weather . . . we have been unable to sail, the winds being S and the sea very rough. At 11 am there was an earthquake which we felt on both the 2 vessels . . . On the shore, the sea rose and fell by about 4 feet in height . . . The earthquake was not extensive and caused no appreciable damage . . .

Caro's journal was much the same.

> It has rained almost continuously since we have been in this harbour . . . We are awaiting fine weather to get out . . . At 11 this morning we felt a movement in the ship that went on for approximately a minute which had the same effect as one feels aft when they are dancing on the forecastle . . .

Fesche displayed even greater brevity.

> The officer reported that outside the winds were S, stormy gale and that the sea was very wild, at 10.15 am there was an earthquake.

Nassau-Siegen told us more, but later changed his mind and rewrote his account without any mention of Jeanne.

> On the 21st at around 11 am the effects of an earthquake were felt very distinctly on board our ships . . . The sea rose very high at the time of the quake but fell right away. Dampier says it is a common happening in the country. ~~The sailors discovered on board the *Étoile* a girl disguised in men's clothing who worked as a servant to M. Commerçon. Without suspecting the naturalist of having taken her on for such a painful voyage. I like to accord to her alone all the honour for such a hardy enterprise, forsaking the tranquil occupations of her sex, she had dared to face the fatigue, dangers and all the events one can morally expect on a navigation of this kind. The adventure, I believe, can take its place in the history of famous girls.~~ After taking possession of the island by leaving an inscription as evidence we left this port on 24 July, well supplied with wood and water.

Vivez says much, changes much between different versions and leaves us uncertain of what he means to say.

> ~~If the reader remembers the facts that I have described to him of Cythera, he must believe that there was no great doubt on the decision about the sex of our so-called eunuch.~~ She remained on our ship where it can be appreciated there remained no doubt in anyone's mind after what had happened but since there was no physical evidence, she dealt with the accusation by issuing some challenge to the servants

> who promised her an examination which occurred at the next place of call in spite of two loaded pistols she always carried by way of precaution and which she took good care to show them to impress them. Going ashore, one ~~fine~~ unhappy day ~~eleventh of this month~~ when I do not know what had happened to the pistols, after having gone botanizing her master left her ashore to look for shells ~~when having seized the pistols we visited the barrel and when we came to fire the lock we discovered the touch-hole which removed all the doubts~~ and the servants who were there drying the washing took advantage of the moment and found in her the *concha veneris*, the precious shell they had been seeking for so long. ~~It was in fact a service that we rendered to this girl that we will now call Janeton~~ This examination greatly mortified her but she became more at her ease, no longer compelled to restrain herself to stuff herself with cloths. She ~~remained blushing as a man~~ finished the voyage very pleasantly, having suitors on all sides ~~but we still admit to ignoring the just cause of her metamorphosis~~ who did not lessen her fidelity to her master. She ended up marrying a King's blacksmith at the Isle de France where I left them and have learned since that she was running a very happy household.

I cannot tell when Duclos-Guyot penned his note about Jeanne's discovery but he writes it in the margin of the following note, on 18 July. There is no entry for the 21st or 22nd.

> The weather is fine enough. The *Boudeuse*'s boat has collected water and ours was used to carry wood. There was a lot of rain overnight.

Only later, in Mauritius, would Commerson write his own version of events, indirectly, when he adamantly declared that

Jeanne had 'evaded ambush by wild animals and humans, not without risk to her life and virtue' and had completed her journey 'unharmed and sound, inspired by some divine power'. And finally, he takes a shot at an unnamed 'detractor', who can surely only be Vivez: 'If by chance she obtains glory or fame, the detractor himself may fall into this poet's curse.'

In a modern biography of Jeanne Barret, English lecturer Glynis Ridley claims that, on the basis of these vague journal entries, Jeanne was gang-raped by the other servants at Port Praslin. The story gets even more elaborate with the telling – that she fell pregnant and later had a child in Mauritius. It's almost as if disaster is what automatically befalls women who dare to leave home, to travel to have adventures. They end up raped or dead, or both. Ridley does not consider rape just as an option, but an absolute certainty.

'Historians of the expeditions who argue that these passages imply only an inspection and not a rape,' insists Ridley, 'are projecting onto Baret's story what they wish had happened, as opposed to what so clearly did.'

It is an accusation, surely, that swings both ways.

Ridley's enthusiasm for portraying Jeanne's story as 'a narrative of a lone, powerless, terrified woman, who was violated emotionally, psychologically and physically', is also a case of reading what she wants to read, seeing what she wants to see and ignoring the evidence. As a lecturer in literature, Ridley would be well aware of the power of narrative, of the clichés and tropes in the stories we tell. In the eighteenth and nineteenth centuries, the stories people told about Jeanne were always ones of the faithful and devoted servant, as if she was some kind of dutiful Sancho Panza, Man Friday or virtuous Pamela.

In the twenty-first century, the servant trope is difficult to sustain. As women have become increasingly disinclined to stay at home and accept domestic servitude, we seem to invent new stories to keep them in their place. We gleefully tell tales about the dangers of travelling alone, the risks of random rapists and serial killers. The dead naked body of Jane Doe fills our bookshelves and television screens with countless variants of the same story, as if women and girls were not at greater risk of being killed in their own homes by their own partners and families than by strangers in the street. The prevalence of such controlling morality tales, the perpetuation of these fears and the absence of stories about the woman who defy them is a tenacious and perplexing feature of human culture.

The presumption of Jeanne's rape 250 years post hoc has spread across the internet in the last decade and is now ubiquitous. Having been said, it cannot be unsaid. It's hard to find a popular account of Jeanne Barret that does not mention this as if it were fact. The loyal, faithful, self-sacrificing servant of the nineteenth century has disappeared, replaced by a victim of the male patriarchy. Both stories are clichés and neither is well-supported by the evidence. There's no denying that women have been oppressed, subjugated and sexually abused for centuries, if not millennia, and that this problem continues today. There are plenty of stories we should tell about this, but surely we don't need to turn every woman's story into the same one. Do we really have to keep figuratively burning every Jeanne d'Arc at the stake for her heresy?

No-one can say if Jeanne was raped or not, in New Ireland or elsewhere. There is no evidence that she was, no proof that she wasn't. But in a society where women were often labelled as promiscuous when they were victims of male sexual violence, Jeanne's virtue and faithfulness were stressed by every single one of the men who documented her presence – even Vivez when he very clearly seems to have wished it were otherwise.

Rape was punishable by death in the French navy. Failing to declare a pregnancy, too, was illegal – also punishable by death, even if this was rarely enforced. By contrast, allowing a woman on board ship was only punishable by a few weeks' arrest or a minor fine. And the French, in the navy as elsewhere, were scrupulous administrators of paperwork. I find it difficult to imagine that if the servant of the King's Naturalist on board a naval vessel had been raped, and/or fallen pregnant, that complaints would not have been lodged, that paperwork would not have been completed, that the perpetrators would not have been punished and that the records would not have been found, or that Jeanne herself, who carried those two pistols in her belt at all times, would not have exacted her own reprisals. At the very least, bitter commentary would surely have slipped into Commerson's private journals and letters.

And finally, what naval commander could afford to overlook such a serious breach of discipline, when both Bougainville and La Giraudais had explicitly given orders to ensure that she was not to suffer any unpleasantness?

You can read whatever you want into this incident. Perhaps I'm reading what I want into it. Many things happened on this journey for which there is no evidence, but history is not the place to discuss them. Personally, I can see no reason why, in an age where women's lives were so constrained, where women were denied so much and permitted to achieve so little, this one woman who managed to escape the constraints of her sex to travel the globe should be a victim instead of a hero.

There is no evidence of a rape. There is no evidence of a pregnancy. But it's clear that Jeanne's true identity, from here on, must well and truly have been known to everyone.

14

STARVATION AND SCURVY

The Spice Islands, August–September 1768

THE HEAVENS OPENED IN one last effort to drown these unwelcome visitors, or at least to wash away their sins. Caught between the north-west and the south-east monsoons, it rains heavily in New Guinea year round, even in the dry season – from 2 to 4 metres, reaching a monstrous 8 metres on some of the clouded mountain tops.

Eventually, towards the end of July, a calm, clear day dawned. Bougainville ordered the tents packed and they departed from this 'wet and unpleasant hole'. He kept the crew busy as they drifted on light breezes away from the shore. The tents were sewn into trousers, for they were not just running short of food, but clothes as well. He could hardly leave his own men as naked and exposed as the locals in all their 'their enormous and pendulous nudity'.

The two ships sailed along the north coast in the stoic hope of reaching their destination before anyone died. The signs of scurvy returned. The sailors were weakened and irritable, their bones ached, their gums bled and their skin bruised red and black.

I am wondering about Jeanne's health. I have read so much about Commerson's and the crew's but, apart from her initial seasickness and Vivez's vague insinuations in South America, there is no record of Jeanne having been ill. She must have had a strong constitution. But they were all on the same rations on the ship, so everyone must have suffered from the same nutritional deficiencies.

'What food, Good Lord, is our lot!' declared Bougainville. 'Rotting bread and small quantities of it and meat the smell of which the most intrepid cannot bear when the salt is removed. In any other circumstances our salted provisions would be thrown overboard.'

Small chance Jeanne could fall pregnant under such conditions. Scurvy is not conducive to fertility. Malnourished and weak, it was unlikely that Jeanne was even fertile towards the end of the journey. In any case, Commerson was a doctor. Both contraception and abortion were so widespread in eighteenth-century France that there was a notable decline in the birth rate in the second half, even if Jean-Pierre Barret's birth in Paris suggests that the former was not always reliable.

All things considered, I am not surprised that Jeanne did not fall pregnant on this voyage. She did not have any children later in life either, despite still being relatively young on her return to France. Ongoing health issues caused by the hardship of a life at sea were common among sailors. Perhaps this was Jeanne's fate as well.

Even the locals did not seem to want to come close to the ships. As they sailed to the north of New Guinea, they saw no-one.

'There have been many arguments about where Hell is situated,' Bougainville said. 'Truly we have found it.'

By September, they had reached the Moluccan islands controlled by the Dutch. Bougainville began to recognise landmarks from his charts, was able to identify bays and mountains and predict settlements. They finally knew where they were and the waters were familiar, even to those who had never been there before. The prospect of resupply and fresh food awaited.

They anchored at a small Dutch posting on the island of Buru and were greeted by two unarmed men, one of whom spoke French. When asked what brought them here, Bougainville replied tersely, 'Hunger.'

The Resident, Hendrik Ouman, sent word to invite the officers to dine with him. When presented with supper they could not help but devour every scrap of it. Their joy at eating fresh meat was indescribable. It felt like the best meal they had ever eaten.

The Resident and his traditionally dressed Ambon wife were kind and generous beyond measure. He hospitably opened his beautiful Eastern home to his unexpected, and somewhat unwelcome, guests. Dutch regulations explicitly forbade foreign visitors. But humanitarian principles overrode that concern – and the fact that the French ships were considerably better armed than the small settlement. They brought Ahutoru ashore, who behaved with all the dignity and decorum one might expect from a Tahitian prince.

It did not take Bougainville long to procure the refreshments they needed to restore the health of the crew. Price was no object. Loaded with rice, dried venison, fish and a few cattle, the ships were restored to some semblance of normality. The sick were sent ashore to revive in the good air and they began to recover from their scurvy, having nothing to do but walk in the attractive countryside and relax.

I don't know if Jeanne accompanied Commerson ashore, where she slept or how she dressed. I expect that she was still dressing as a man, and that she was still assisting Commerson with his work. Commerson's leg wound had reopened in Tahiti and he continued to suffer from it for the next two or three months. He would have needed Jeanne's assistance and was not likely to have received it from anyone else. She may also have felt safer keeping close company and continuing to work and behave as she had before.

The native wildlife was spectacular – birds of wild and ornate plumage, highly collectable shells, cockatoos, a strange wild cat that carried its young in a pouch, a huge bat, and snakes so big they could eat sheep while others flung themselves from trees onto those passing beneath. And then there were the enormous crocodiles that infested the riverbanks and indiscriminately hunted both man and beast.

In the online herbarium collections of the Muséum National d'Histoire Naturelle, it is easy to find the plants that Jeanne and Commerson collected from here. The keywords of 'Commerson' and 'Bouru' or 'Indonesia' swiftly bring to view the beautiful pink dendrobium orchids, black pepper vine, cannas and lilies. Some were collected from the woods, but the lily was from the governor's garden. I don't think they ventured far – it is a fairly minimal collection from a country full of botanical beauties, oddities and treasures.

They also collected birds, including a spectacular rainbow bee-eater and kingfisher. But some birds are anomalous. Not all of them should have been here. Commerson's collection also includes a southern cassowary, not found in the areas they visited, and birds from the Philippines. It is possible Commerson obtained these later, by trade in Java, or in Mauritius from his colleagues Pierre Poivre and Pierre Sonnerat, who had been to the Philippines.

In Commerson's scrawled notes on birds, there is also a description of *Psittacula brevicauda*, which literally means a short-tailed parakeet. Commerson's handwriting is too hard for me to transcribe and translate from Latin, but he mentions green a few times, as well as red, yellow and black. A colourful parrot with a short tail then. I cannot tell which of the twelve parrots found on Buru this description might refer to. But in the margin Commerson has written, '*le perruche de Baret de Bouru, Septembre 1768*'. He wasn't naming it after Jeanne but perhaps he was noting that she caught it? I don't know, but it is one of the few times Commerson ever wrote Jeanne's name.

I had reached the end of my voyaging at Madang in New Guinea. It was my only ocean crossing and my last time living aboard *Caliph*. I flew back to Australia to continue my studies while Dad sailed the boat back some time later.

We had already stopped living on the boat by the time we reached north Queensland. My Siamese cat, my constant companion since I was four years old, who had bested dogs, snakes and sharks, been lost overboard, chased by a shark, abducted, left behind in port and yet followed us everywhere by car, boat or on foot, retired comfortably into a terrestrial old age, while I headed off to study at university on the other side of the country.

It had taken us four years to build the boat, a year to refine it, and three years to travel from South Australia to the far north. It was neither a long journey nor a particularly adventurous one, but it was rich in memories and experience. It left me with a passion for wilderness, for the diversity of the Australian landscape, and an unquenchable thirst to test the assumptions behind the stories we tell and the history we think we know. Not all voyages end just because you step ashore.

They would have liked to have stayed in Buru for longer. But the end of the eastern monsoon was approaching and with it, contrary winds and currents. No matter how good the food, they had to leave, and quickly. The English, in any case, were up to something. Their activity was everywhere. A ship conducting a world tour had left just eight days before their arrival. The Dutch would not maintain their monopoly on this spice trade for long, it seemed.

They navigated the shoals and shallows of the Macassan Straits with more ease than the Dutch claimed possible. There were plenty of locals keen to sell them food, no doubt at a premium but still cheaper than from the Dutch, and both ships were soon full to the gunwales with fruit and fowl, as well as a prodigious quantity of cockatoos and parrots with the most beautiful plumage. With all the fresh food, their scurvy abated, but they knew, even as they approached Batavia (now known as Jakarta) on the north coast of west Java, that it was the start of the notorious fever season.

The city was a wonder, simultaneously exotic and cultured. How could they tire of the wondrous feasts, the exceptional company, the Chinese concerts and comedies, the charming walks, the conspicuous wealth and luxury, the monuments and grand mansions and gardens of the city, so fine that they rivalled even Paris? The officers and passengers, and presumably their servants with them, stayed in a hotel in the city and hired several coaches to take them wherever they needed to go. While Bougainville and his officers made all the required courtesy visits, Commerson and Jeanne attended to their collections, which swelled here. Whether they travelled to the outskirts of the city to collect for themselves, or simply traded specimens with other interested naturalists, is impossible to tell. Batavia was a centre for all trade in the region – including natural history and conchology.

Even so, I am unlikely to have much luck finding any shells that Jeanne collected here. The Portuguese, the Spanish and the Dutch had been bringing Pacific shells back to Europe for decades, and the shells from this region were already well known in Europe. The Dutch apothecary and collector Albertus Seba published a pictorial inventory of his natural history collection between 1734 and 1765. It was an impressively cosmopolitan collection, including a great many species from the southern and Pacific hemispheres. The book contains one of the earliest depictions of a marsupial 'opossum' and a fine illustration of a sulfur-crested cockatoo. It covers a full taxonomic spread, as well as occasional fantasy animals, bottled foetuses, two-headed creatures and minerals. And among the many plates of shells are species from the Indo-Pacific. There are no fewer than seven plates illustrating cones – around 230 or more than a quarter of known conid species. Not bad for a collection of shells from the other side of the world in the early 1700s. I increasingly feel that in hoping to find Jeanne's shell collection, I am chasing a mirage.

The city was insufferably hot. The canals created an unwholesome humidity and the drinking water was so bad that the wealthy Batavians would drink only imported Seltzer water. And there was something about the extreme privilege and pretension, and the harsh duress under which the general population lived, not to mention the slaves, that made even the French uncomfortable.

'We had scarce been above eight or ten days at Batavia, when the diseases began to make their appearance,' said Bougainville, hastening his plans for departure. Ahutoru fell gravely ill here, swiftly falling into docile submission from the excitement with which he had greeted everything new and wondrous. He recovered under careful medical care, but thereafter referred to Java as *enoua maté* – the land that kills. Two of his countrymen who later sailed with Cook on the *Endeavour* died here.

Compared to other long-distance expeditions struggling against tropical fevers and scurvy, Bougainville did remarkably well. He knew the importance of fresh food and good shipboard hygiene. He understood the risks of bad water, poor food and diseased locations. He prioritised the health of his crew and adjusted his voyage to minimise those risks and prevent deaths. He lost only seven lives out of well over 340 men on the journey – just 3 per cent. It was an excellent achievement at a time when earlier long voyages had lost up to half of their crew to scurvy and other illnesses. Bougainville lost fewer men than the contemporary expeditions of Byron or Wallis and far fewer than Cook, who lost more than 30 men – one-third of his crew – to dysentery and typhoid on his first voyage in the *Endeavour*, which left shortly before Bougainville's return.

Everyone must surely have been tiring of their journey by this time, including Jeanne. The revelations about her identity would almost certainly have changed the way people behaved towards her. I suspect she lost something of the freedom she had enjoyed when she first embarked on her journey, the freedom many women report when they take up a male persona: the freedom to be appraised as an individual, rather than either as an object of sexual desire or an object of no importance whatsoever. Commerson may not have cared who people were, a servant or an aristocrat, a slave or a savant, provided they were intelligent and had an interest in natural history. He may not have particularly cared if Jeanne was a woman or not, but it is certain that most other men did. Even if they sometimes forgot, and reverted to treating her as they had for so many months on the journey as a man, there would have been times when they remembered, and their expressions would have changed, their friendship or tolerance been withdrawn and replaced with either a new reserve or an entirely unwanted interest.

Bougainville's haste to leave was well judged. Without waiting even to load further livestock they left in mid-October, having by and large completed their mission's objectives, and headed for French territory in the Indian Ocean, the distant speck of Mauritius, then known as Île de France.

In these familiar waters, the *Boudeuse* dispensed with keeping pace with the *Étoile* and promptly disappeared over the horizon. They were on the home strait.

There is nothing to do on a long sea voyage. In the steady winds, the sails set so still and unmoving that you sometimes have to tap them to check that they have not turned to stone. You walk the deck to relieve the boredom but before you know it you are back where you started. Nothing has changed. The sea still circles in a never-ending circumference as if you are making no progress at all.

You stand at the heads to watch the bow wave surging beneath your feet. Or stand at the stern and watch the ship's wake foaming with glowing phosphorescence into the far distance. It gives some sense of movement. You can't help but think about what would happen if you tripped and tumbled into the watery depths. If anyone would even notice if you were gone. If the ship would just sail on without you until there was nothing at all but you and the sea and the sky all around, all the way to the heavens and all the way to the darkest depths.

Imagine that first shock, the gasp of breath, the lungful of water, the strangled struggle to breath, to float. Control the panic, still the flailing arms and legs, chain them into dutiful strokes, calm the wheezing gulps of air. You are afloat and you can breathe. The water is the same temperature as your blood. You can barely tell it's there – that there is a difference between air and water.

Imagine sinking. This is a fantasy – you don't need to breathe. Let the water rise up over your mouth, over your eyes. Relax. Let the water fill your lungs like oxygen-rich amniotic fluid. They've done this before. They'll remember what to do.

You are sinking slowly through the euphotic zone, the sunlit region of the upper oceans. Here the water sparkles with phytoplankton and marine plants charging themselves on the solar power of the sun. The water is clear, mid-ocean, free from the turbulence and debris of the coastal fringes. The sun's rays reach deep, more than 200 metres, before they become too weak for photosynthesis.

Beyond this lies the twilight zone – the bathyal layer, where light still filters dimly but photosynthesis is no longer possible – and then the deep-sea zone where no light reaches at all. This is the cemetery of the ocean, where the life of the surface sinks to its end. Despite the darkness, the lack of plants, life thrives. The detrivores rule, feasting on the bounty that falls from above. But there are other sources of energy here too. In the cracks on the ocean floor, along the Pacific, Southeast Indian and Central Indian Ridges, deep-sea hydrothermal vents spurt geysers of heated water into the frigid depths, crystallising minerals into smoking spires and chimneys. Chemotrophic bacteria feast on hydrogen sulfide and methane expelled from beneath the rock, while 2-metre-long tubeworms provide forests for tiny fish, crabs, shrimps, sea snails, mussels and limpets. It is an inconceivable world filled with unimaginable creatures.

What is it about the sea that inspires such dread imaginings, such a sense of loss and irrelevance, of floating in empty space? Perhaps is it that we are drawn to the earth, that we are terrestrial, not oceanic, pelagic, demersal or aerial? Our two feet long to be planted firmly on the ground. Gravity pulls us down, even against the pressure of the water, against the dark horrors of the depths, the water in our lungs and the cemetery of skeletons and zombie worms that await in Davy Jones's locker. An alien world in which shallow-lunged air-breathers like us have no place.

But we are nothing if not explorers. No alien world has been forbidden. We are never content to stay home where we belong. We streamed north out of the cradle of Africa not once, but multiple times, waves of human emigration like a tide rising on a beach, each successive wave outreaching the last until eventually we had spread throughout Europe, across Asia and down into Australia. But still the waves of people kept coming, diluting, intermingling, changing and adapting. Sea levels changed, land bridges disappeared, populations were isolated. The ice retreated, the seas fell, and the Bering land bridge opened the floodgates into the Americas, the human flow cascading down the continent to reach the tip of Tierra del Fuego at least 12,000 years ago.

Neither frozen wastes nor searing deserts deter us: no jungle is thick enough to withstand penetration; our trail is blazed with the bodies of our explorers and those who stood in their way. No continent remains untrodden. Our footprints dent the lunar landscape and our trails of rubbish spiral out in a dirty orbit from our planet, matching the spirals of rubbish that circle and collect at the heart of the Pacific, as even more of our refuse drifts, unheeded and unnoticed, into the depths below.

It is hard to predict the value, or even the consequences, of a discovery. Who can predict the lives saved from antibiotics from the fungus growing in a lab during a break? It takes a special kind of mind to foresee microwave ovens from a chocolate bar melting near a magnetron. Or Velcro while picking burrs from your socks. Or frequency-hopping torpedos and wi-fi from player pianos.

Of what value is a voyage of discovery through the Pacific? Perhaps it should be measured by the new lands it finds, the coastlines it charts, the colonies it founds, the territory it claims, the species it identifies, the collections it brings back or the riches it steals. Or should it be measured by the cost, the diseases spread, the species introduced, the violence done and the cultures destroyed?

'It is easy, no doubt, to exclaim,' Commerson wrote to his brother-in-law, 'as is already done on all sides, that the journey is beautiful! That it brings glory! But who can imagine its cost? A thousand pitfalls faced by night as by day, the foulest and most revolting food, the dogs, the rats, the leathers of our ships prepared by the hand of famine which pursued us for several months; scurvy, dysentery, putrid fevers carrying off the best of our troop, and, sadder still, a state of distrust and internal warfare pitting us against each other. These are the shadows of this grand and beautiful history.'

Saint-Germain, the clerk, was more succinct.

'Of what use is this voyage to the nation?' he queried peevishly.

15

JOURNEY'S END

Mauritius, November 1768–1770

The *Boudeuse* arrived off Mauritius in early November 1768, just as night fell. They fired their guns to request a light on Gunners Point to guide their safe passage in through the shoals. But there was no response.

They picked their way carefully around the offshore reef and tacked back and forth off the port through the night, firing their guns at intervals, until finally, near midnight, a pilot came to guide them – and promptly drove the ship aground in the shallows of the Bay of Tombs.

Bougainville fumed as he ordered the sails backed while the crew prepared to bail and lower the boats. The ship heaved itself off the shoal, stripping off most of the false keel in the misadventure.

'What a fate it would have been to come aground in port through the fault of some ignoramus paid by the king and for having followed the rules!' the commander railed.

The *Étoile* arrived the following night and more cautiously stood offshore, waiting until daybreak to approach.

They had finally reached home soil, a little outpost of France in the middle of the Indian Ocean, where they would greet old friends, hear news from home, eat their own food and speak their own language. For many it was the end of their long journey.

And for many who have told Jeanne's story over the years, Mauritius was the end of hers too: a place where she disappeared, died or simply faded from their interest, as if her life was never anything more than a minor accessory to the story of more famous men.

I have never approached Mauritius by sea, so I can only imagine how its sharp distinctive mountains might have risen from the water to welcome weary sailors back to a distant garrison of France. It is striking enough to arrive by air. As my plane tilts on its final descent, there is a sharp gasp from the passengers as we catch a glimpse of our destination in the late-afternoon sun. Angular volcanic peaks jut almost vertical from a green plateau, ringed by glistening white beaches in a sea of the most astonishing blue. Ripples of reef enclose viridian bays protected from oceanic breakers. In its pristine isolation in the middle of the Indian Ocean, Mauritius presents a picture-postcard image of idyllic tropical beauty.

By the time we land, night has fallen with sudden equatorial surety. The dark warmth gusts through the open-plan airport, redolent with the earthy aromas of fecund humidity and decomposition. Tourists struggle to pile overloaded suitcases into taxis, on their way to beachside hotels and villas, and trailing the scent of coconut sunscreen and holiday indulgence. I lower the window of the taxi, enjoying the warmth. As we slip between rows of thick vegetation, I catch glimpses of vast sugarcane fields through tangled regrowth of guava, eucalypt, lantana and privet. Weeds, I think instinctively. I can't help myself. Tropical paradise or not, Mauritius is famous among conservationists for reasons other than its beauty. It is an island synonymous with extinction. 'Dead as a dodo,' as they say. The state symbol of Mauritius is a bird that went extinct so long ago that we barely even know what it looked like.

I'm not travelling to Mauritius to research Jeanne's life explicitly. My research is never so convenient and linear. I am attending a conference on climate change, and I can hardly resist the temptation to visit the island that was the centre for so many of the French voyages of discovery to the Pacific. Such research always comes in handy one day.

It's one thing to fly for eight hours across the Indian Ocean to a research location, but it's quite another to strip back the layers of history, undo centuries of development, revegetate the mountains, learn to recognise the differences between British and French colonialism, and try to see the traces of the past written into the landscape, hidden in the backstreets, in museums and landmarks.

I can barely match the modern Port Louis to the one in the engravings from the late 1700s. I can see the same layout of streets, the same shape of the harbour, but I can't reconcile the concrete docks, cranes and steel-hulled supertankers with the port that the *Étoile* and *Boudeuse* entered, scattered with shipwrecks and debris washed down from the mountains. This redeveloped foreshore filled with high-rise hotels bears no resemblance to the view that Jeanne would have had as she arrived. I take my bearings from the Jardin de la Compagnie,

Petite Montagne and the angular mountain Le Pouce that rises behind, but I quickly get lost in a labyrinth of alleyways and side streets that feel more Asian than European, more European than African, and more African than Oceanic.

Port Louis is nothing like 'the Camp' Bernardin de Saint-Pierre described when he arrived in Mauritius a few months before Jeanne and Commerson. He thought that the capital was located in the 'most disagreeable part' of the island.

'The town,' he wrote, 'has scarcely the appearance of a market town, is built at the bottom of the port, and at the opening of a valley . . . formed by a chain of high mountains, covered with rocks; but without trees or bushes. The sides of these mountains are covered six months in the year with a burning herb, which makes the country appear black, like a colliery. The edge of the rocks, which form this dismal vale, is broken and craggy . . . As for the town or camp, it consists of wooden houses of one story high; each house stands by itself and is enclosed in palisades. The streets are regular enough, but are neither paved nor planted with trees. The ground everywhere so covered, and as it were flaked with rocks, that there is no stirring without danger of breaking one's neck.'

Perhaps it was not so disagreeable to the expedition members, who vacated their ships like proverbial rats. Everyone wanted to go ashore: to feel firm land beneath their rubbery sea legs; to breath fresh air and taste fresh water; to eat fresh meat and vegetables that had not been salted, stewed, dried or putrefied; to drown their sorrows and forget their hardships in bars and brothels. Only enough crew stayed on board to tend to the needs of the ships.

One hundred and sixty-one men were sent to the hospital to recuperate. Not all of them would return. The fevers of the tropics had not yet claimed all their victims. And the ships, too, needed tender care. A sheltered place behind the causeway linking Tonneliers Island to the mainland provided just the right slope for vessels to nudge themselves ashore on a falling tide – with barely an hour or two to strip off their verdant slimy undergrowth as it dried and stank in the heat, before the sea rose and tipped the ship to the other side. Rinse and repeat.

As the smallest person on *Caliph*, I remember crawling beneath the hull to clean near the keel, a task I completed neither efficiently nor enthusiastically. It was my second-least favourite chore of childhood, surpassed only by hanging upside down in the chain locker to stow the wet, muddy, finger-crushing anchor chain and narrowly beating pumping out the bilges. So much for the romance of the sea.

The *Boudeuse*'s sheathing was worm-eaten and the masts needed repair, but it was otherwise sound. Bougainville transferred all the healthy sailors to his ship and prepared to depart. The *Étoile* would need a good month longer in harbour to make all the necessary repairs. There was no need for the ships to travel together on this homeward stretch.

On the eve of his departure, Bougainville tallied up the gains and losses to his crew.

'I left behind for the King's service in the colony,' he recorded, 'the Reverend Father Lavaisse, chaplain; Messrs Fetch, volunteer; Verron, pilot observer; Oury and Oger, first and second pilots; De Romainville, infantry lieutenant sailing in the *Étoile*; Pierre Duclos' son, volunteer ditto; Commerçon des Humbert, naturalist ditto and his valet, a girl as a man.'

Commerson and his 'girl as a man' would have been met by Pierre Poivre, the intendant or administrator of the island. Poivre was a handsome and charismatic man by all accounts – tall, good-humoured and with a profound knowledge of, and passion for, natural history. It was Poivre who would finally break the Dutch monopoly on the spice trade, establishing nutmeg, pepper, cinnamon, star anise and cloves as well as avocados, mangoes and cocoa crops in the French colonies.

Poivre and Commerson were friends. They met in 1758, when they both lived in the Lyon area. They met again in Paris in 1766, when they were both preparing for their respective departures: Commerson, with Jeanne, on their expedition around the world; and Poivre, with his new wife Françoise Robin, to take up his post as the administrator of Mauritius. It is likely that Poivre would have met Jeanne in Paris, or at least have known of her contribution to Commerson's work. Whether he knew she would be travelling with Commerson is uncertain, but he would have known who she was when she arrived in Mauritius.

Poivre knew Commerson was coming to Mauritius and he desperately needed a botanist to help him with his garden at Pamplemousses and to document the potential of the native flora. Bougainville, preparing to leave for France, no longer had any need of Commerson's services.

It is tempting to imagine that Bougainville disembarked Commerson and Jeanne at Mauritius to avoid any controversy when he returned home. If and when they returned to France their conduct would no longer be Bougainville's problem. But Bougainville had never seemed particularly concerned by the illegality of Jeanne's actions.

'The Court, I believe, will forgive him for breaking the ordinances,' he said.

In any case, a great many people transferred from ship to shore at Mauritius. Bougainville replaced all his sickly sailors with healthy ones ready to make the trip back to France. He left behind all his expeditionary stores, a platoon of soldiers, and all the building materials and medicines he could spare. Both his scientists, Véron and Commerson, stayed on to complete their research. Véron wished to track the transit of Venus in India. Commerson had his eye on Madagascar. There was still much work to be done.

For many years I puzzled over the origin of the strange little broken rings that so often washed ashore on our beaches. They looked like some kind of plumbing fitting – a flange or washer – moulded out of 'plasticised' sand. It took me some time to discover that these collars of sand encased the eggs of the predatory moon snail, which is also responsible for drilling holes in cockle shells.

The egg cases of molluscs are often washed up on the shore, mysterious and diverse. Not all of them are as elaborate, or as carefully cared for, as that of the paper nautilus. Some are plain and utilitarian. Others are elegant, spiralled, symmetrical, ribbed, fluted, coiled, filamentous and beaded – like some space-age alien's medical equipment. They are as diverse as the creatures they grow into.

Some shells guard their eggs, others equip them with robust housing and leave them to their own devices. Some of the eggs hatch into tiny shells, miniature adults, but more often in a marine environment they hatch as free-floating larvae. These trochophores metamorphose into variously shaped veliger phases, hitchhiking on the ocean currents or on fish, to disperse before finally settling – and emerging butterfly-like – into their final form. Some stick themselves to the rocks near where they were born and resist all efforts to be dislodged. Others readily let go and drift, confident that they will land somewhere else where they can be equally at home.

Not all of us know what our final form might be. It takes a while to find the right place, the right life, to settle into.

There is something about the tropics – that rich abundance of biodiversity, the lush forests and fertile soils, the speed of growth – that inspires both admiration and complacency. The garden surrounding our house in north Queensland was filled with both exotic and native wonders. Green tree frogs took up residence in the toilet and the kettle. Geckos and bright jumping spiders patrolled the ceilings. Happy plants that sold for $50 a foot in Sydney soared up to my first-floor window, filling the night air with their pungent purple perfume. A giant raintree cast its lacy umbrella over the backyard and house. Orange starbursts of epiphytic native orchids wove through the shrubbery. Delicate ferns with iridescent shades of blue and pink coated the bank. Mowing the grass was a weekly task simply to avoid a jungle.

My parents took to the overgrown garden with vigour, stripping the ferns back to bare earth and removing the happy plants from my window to make way for new stumps under the house. But the garden did not recover as quickly as we thought it would. The red earth baked hard beneath the unaccustomed sunlight, soil washing in deeply carved rivulets down the bank and onto the driveway. Across the river, in the farmland and canefields, you could see the same pattern exacerbated – land stripped bare and unable to regenerate in the unforgiving conditions. Cleared land cut into forest, which sealed their open wounds with vigorous climbers and invasive weeds. Small wounds can heal, as our garden did, but larger gouges leave openings for invasion, infestation, infection and disruption.

My classmates did not see always this. They had grown up surrounded by vigorous healthy forests that seemed insurmountable and inaccessible. They had so much water that they routinely left hoses running full bore into the gutters. Few of them had travelled – even Brisbane was the 'far south'. Some of them had never crossed the ranges inland to the desert country. The notion of environmental protection was anathema to many of them.

One of the maths teachers spoke to our geography class about conservation and the threats to the nearby Daintree Rainforest. He spoke quietly, as if confessing a sin. He asked us not to mention it to anyone for fear of losing his job or suffering reprisals. He sweated profusely and his hands shook as he spoke.

The Daintree National Park should have been safe from development, particularly along the coast where the Great Barrier Reef adjoins the land. But the local council wanted to push a road through to open up access to the north. Despite vocal opposition, the roadworks began in 1983. Predictably, the wet season halted construction for six months as landslides deposited mountains of silt onto the World Heritage–listed Great Barrier Reef.

How easily such riches are lost and despoiled, sacrificed to human growth, expansion and survival. The road is sealed now, and few would even remember the thoughtless damage done by its construction. But the reef itself continues to suffer from a thousand cuts – run-off from agricultural and industrial developments, or overfishing, or dumping from dredging, or broadscale coral bleaching from an ever-warming climate. The great reef may well be beyond saving now.

We have lost our reef-builders in every mass extinction event the earth has witnessed. They are always the first to go, and each time it has taken over 100 million years for new species to take their place. They do not recover quickly and there is no reason to think they will this time either. They will join the list of other human-caused extinctions, which is so much larger on long-isolated land masses like Australia and Mauritius.

Isolated for millions of years, the rich endemic plant and animal life of Mauritius diversified and evolved almost without any mammals – without humans. Fruit bats were the only mammalian colonists, and bird life flourished – much of it unique to the island.

Despite lying just off the coast of Africa, both Madagascar and Mauritius were isolated from human contact until relatively recently. Remarkably, Madagascar was first settled, not by Africans a short distance from the west, but by Austronesians on outrigger canoes from Borneo, in about 550–350 BCE. The tiny pinprick island of Mauritius took even longer for humans to find. It was depicted on Arab maps in 1502, and by 1507–13 the first Portuguese sailors visited the area, giving the Mascarene island group its name. In 1598 an expedition of Dutch ships on their way to Indonesia landed here after bad weather, naming the island Mauritius after one of their ships, and establishing the location as a safe stopover. Like many introduced species, establishment of a viable human population on Mauritius took effort and persistence. For more than a century, the early Dutch settlements struggled to survive in the face of cyclones, illness and starvation, yet left a considerable legacy. By the time the Dutch abandoned the island, they had introduced sugar cane and rats, eradicated the dodos and giant tortoises, and cleared large swathes of forests of ebony.

The French moved quickly to claim the abandoned colony in 1715, and the arrival of the Breton governor Bertrand-François Mahé de la Bourdonnais in 1735 saw Port Louis successfully established as an administrative and port facility with the foundations of a viable agricultural industry.

By the time Jeanne and Commerson arrived on Mauritius, the long fingers of deforestation were already spreading up the valleys from the main coastal settlements into the hinterland. Within 50 years, when the English had taken over the island, almost half of the land was cleared, and within a century less than a quarter of the native vegetation remained. Today less than 2 per cent survives. The rest is cultivated for agriculture or covered with a mongrel mix of introduced environmental weeds.

It's hard to even imagine the ebony forests for which Mauritius was once famous. Ebony provided the highly prized black timber for piano keys, furniture and jewellery. The largest trees were thousands of years old, their stocks soon exhausted by harvesting. Today, the remaining protected forests are dominated by small trees and harvesting is no longer possible. Almost a third of the island's endemic plant species are critically endangered, some represented by just a handful of known specimens.

In the 400 years since human settlement, over 100 plant and animal species have disappeared, including two giant tortoise species, a giant skink and two fruit bats, thirteen birds, and at least thirteen endemic snails. Like the dodo, many went extinct before even being described – the early victims of rats from ships, as well as introduced cats, mongooses and monkeys.

It's impossible to protect things that you don't understand, things you don't even have names for. Commerson and Jeanne's work could have made a huge contribution to protecting and valuing the forests and the animals they supported. The palms alone were worthy of their own book, or perhaps a volume in a series on the natural history of the islands. What difference might an early publication on the unique species of these islands have had? We will never know.

While Jeanne and Commerson settled into their new land-based life, the *Boudeuse* headed back to France, arriving at Saint-Malo on 16 March 1769 and thus completing France's first successful circumnavigation of the globe. And when Ahutoru stepped onto the docks at Saint-Malo, he became the first Pacific Islander to discover France. Unlike those who travelled to his country and based their observations on a brief and bloody stay, Ahutoru was an attentive and diplomatic guest when he arrived in Paris, despite his lack of French and his host's lack of Tahitian.

'He went out by himself every day, and passed through the whole city without once missing or losing his way,' said Bougainville. 'He often made some purchases, and scarce ever paid for things beyond their real value. The only show which pleased him was the opera, as he was extremely fond of dancing. He knew perfectly well on what days this kind of entertainment was played; he went to it by himself, paid at the door the same as everybody else.'

Furthermore, he made friends – and seemed to value the gift of their kindness and concern much more than he did any material offerings.

'Were these Savages?' the generally pragmatic Fesche had asked of the Tahitians. 'Certainly not, on the contrary we were the ones who had behaved like barbarians and they acted like gentle, humane and well-regulated people. We murdered them and they did only good to us.'

Bougainville, meanwhile, was kept busy publishing his account of the journey. Many considered the voyage a success, and Bougainville would go on to have a long and influential career in the navy, sciences and political diplomacy. He had completed the first French circumnavigation, with the loss of only seven lives. The public appetite for travel narratives was strong, their value as documents of national status was well appreciated, and the importance of publishing his own account before any of his other officers could was pressing. If Bougainville needed convincing of the urgency, a letter sent to the French Academy of Sciences, and published in the *Mercure de France* in November 1769, would only have reinforced his haste.

The letter had been sent by Commerson to his friend Jérôme Lalande shortly after they arrived in Mauritius, who forwarded it on for publication. His 'Postscript: On the Island of New Cythera or Tahiti' (as Ahutoru called it) was a sensation. A land 'without vices', recognising 'no other god but love' and populated with beautiful sisters of the 'utterly naked Graces' without shame or modesty. Perhaps it is fortunate that Jeanne's family were unlikely to be reading this publication.

European imaginations required no further prompting to adopt the notion of the 'noble savage' and extol the virtues of a state of nature, for all Ahutoru assured them that reality did not live up to their Romantic notions.

Bougainville's account of the voyage was published 1771, and was quickly translated into English by George Forster, who travelled with Cook on his first circumnavigation, thereby giving him the authority to contribute numerous quibbling and often irritating commentaries in the footnotes of Bougainville's account. As Bougainville himself noted acerbically in his introduction:

'I am a voyager and a seaman; that is, a liar and a stupid fellow, in the eyes of that class of indolent haughty writers, who in their closets reason *ad infinitum* on the world and its inhabitants, and with an air of superiority, confine nature with the limits of their own invention.'

It is hard to know what impact the publications about the voyages in France might have had, if any, on the lives of Jeanne and Commerson. But in the long run, the publication of Bougainville's narrative publicly identified Jeanne as the first woman to sail around the world. And later, Denis Diderot's idealistic and fictionalised *Supplement to Bougainville's Voyage*, written in 1772 but not published until 1792, irrevocably connected Jeanne in the public imagination with Commerson and his accounts of Tahiti's free and easy life of pleasure.

~

If Jeanne Barret had been with child from a rape in New Ireland, she would have been three and a half months pregnant on her arrival in Mauritius.

Even though I know there is no evidence of a rape and that these events are implausible and unlikely, I still feel obliged to search for proof. Perhaps this is the curse of the scientist – rather than searching for evidence that supports our theories, we compulsively search for evidence that will prove them wrong.

So I keep looking for a child born in the middle of April of 1769. Barret is and was a common name on the island. But I can find no records of Barrets born in 1769 or the 1770s in Mauritius. I search through a website of Catholic parish registers transcribed by volunteers, but there are no records of any births under Barret, Bonnefoy or Commerson, or any myriad alternative spellings.

That's not to say they weren't there. I know that records exist that are missing from this list. An absence of evidence is not proof of anything. Some proof might turn up in the future. But I can find nothing to support an already implausible argument.

When the *Étoile* finally departed for France, Jeanne's connection with the expedition ended. If Commerson was no longer the King's Naturalist on the Bougainville expedition, was he still employed by the crown, could he still draw his 2000 livres per annum salary and could Jeanne still be paid a measly twelve livres per month to be his servant?

By some wrangling, Poivre managed to gain approval for Commerson's expenses to be transferred, so that he received 3000 livres for salary and housing plus payment for an illustrator at 1800 livres. There is no mention of a servant's salary, although Commerson's own budget would have been more than enough to accommodate that small expense if he chose.

Biographers differ in their accounts of where Commerson lived. Some say Poivre installed Commerson and Jeanne in a large house next to his own. But Commerson's brother-in-law, Father Beau, recalled differently.

'Poivre, who was then intendant of the Isle of France, declared himself the zealous patron of this scholar,' said Father Beau. 'This generous protector gave him an apartment in his house large enough to be able to prepare and conserve the plants, birds, insects, shells and quadrupeds for Commerson's collection. For almost four years he gave him his table, servants to serve him and finally very generously provided him with all the services necessary to make his talents productive.'

The Poivre residence was Château Mon Plaisir in the Pamplemousses Gardens just outside town. Was this where Commerson lived, with Jeanne in the servants' quarters? It seems an idyllic location for a botanist, within a botanic garden, watching over the progress of the seeds and plants he brought back for his friend as they germinated and grew.

The garden at the time was filled with cinnamon trees, palms, a varnish tree and the only specimen of Tahitian mango tree to have escaped its homeland – which can only have been brought by seed or as a sapling by Commerson. There were 'a multitude of trees and shrubs arranged in the finest order', plants both rare and useful according to Bernardin de Saint-Pierre.

'A brook circulates and maintains the freshness of these charming places: the bamboo alleys that surround it, and which resemble from afar our willows, the beauty of the plain and hills dotted here with houses and groves, and even the neighbouring church and bell tower add to the pleasure of the landscape: it gives it an air of France.'

Today, the Pamplemousses Gardens are cool and shady, popular with visitors and locals alike. The carpark is full and buses regularly disgorge restless passengers who tumble hurriedly into the gardens, before slowing and dispersing under the meditative influence of their surroundings. A sign exhorts me to keep to the path, while another warns not to walk under the trees. All the paths are lined with trees. The fierce biting sun persuades me to ignore the risk of falling branches.

I walk under vast twisted fig trees that guard the gate towards the open lawns featuring an arboretum of trees, decorative shrubs and elegant palms. The Tahitian breadfruit tree filters light through its deeply lobed leaves. I admire the pink and white flowers fallen from a tree before I realise that it is the box fruit or sea poison tree that Bougainville wanted to name *bonnet carré* after the square medical graduate's cap that Commerson would have worn.

A row of young saplings has been planted along a path by various dignitaries, commemorating the significant plants of the island, both native and introduced. A nearby marble column is inscribed with familiar names, beginning with the founding governor Labourdonnais, who first purchased Mon Plaisir as a vegetable garden for the colony, then Poivre, Cossigny and Commerson. Other familiar names, from other stories, are scattered below – Rochon, Baudin, Lahaie. Nearby, a bust of Pierre Poivre, with tip-tilted nose, smiles cheerily at visitors. He looks the perfect model for the Peter Pepper who picked a peck of pickled peppers from beneath Dutch noses to grow in his garden.

Although the garden began with Labourdonnais, it is Poivre who is regarded as its founding director, the one who transformed it from a utilitarian vegetable and flower garden into the first botanical garden in the tropics. He restored the neglected Château Mon Plaisir, which still stands in the heart of the Pamplemousses Gardens, as a home for his wife and children.

Mon Plaisir is a simple but elegant 'plantation-style' building: two storeys with wide verandahs all around, a simple slate roof and multiple large French doors from every high-ceilinged room. Unlike much colonial English architecture, this early French colonial approach adapted swiftly to the climate. The building is closed when I visit, but the shutters on the French doors are open, allowing a glimpse of light-filled entrance rooms with glossy timberwork and a carved staircase leading to the upper floors. It would have been a comfortable, but by no means spacious, home. I cannot see how Commerson's collections would have made convivial housemates in this modest chateau. As Lalande recalled, 'it was difficult to find shelter elsewhere with the prodigious clutter of his collections, and the kind of infection caused by his plants and fish, an unbearable smell for anyone who did not share his passion for natural history'.

My memory recalls to me the faintly fetid aroma of my cabin's cupboards as a child, reminiscent of garden fertiliser, which came from the more freshly prepared shell specimens. I did not mind the smell. It was nothing compared to the stench of actually cleaning the specimens that had been hung in the sun for several days sealed in a plastic bag. I held my breath to avoid gagging as I drained the toxic black liquid over the side before rinsing them in a bucket. It was a smell that lingered.

But for all I like the idea of living in Mon Plaisir, I soon realise that neither Pierre Poivre nor Commerson, nor Jeanne, lived there.

'I hardly ever enjoy M. Poivre's company,' wrote his wife pointedly to Bernardin de Saint-Pierre. 'I would be charmed to talk to him a little if he comes on Sunday, and I would not be sorry if he came alone.'

While his wife and family enjoyed the more pleasant and healthy climate of the country, a 12-kilometre horseride from town, Poivre's own archives reveal that he lived in the Hôtel de l'Intendance, opposite Government House in the centre of Port Louis. It was in this more spacious building that Commerson also lived and worked with all his collections.

But I don't know where Jeanne lived, or in what guise. As either a man or a woman, she seems to have slipped into the fabric of island society almost without notice. I don't think she would have stayed in her masculine guise and continued her work with Commerson. Commerson was now Poivre's guest and Poivre knew that Jeanne was a woman. I wonder if the intendent of the colony could afford to be associated with such scandalous behaviour. And so I assume she transformed, on her return to French society, back into a woman. Either way, her presence certainly did not seem to cause any great fuss. Neither Poivre nor Bernardin de Saint-Pierre mentioned any scandal when Bougainville's ships arrived. There was hardly any mention of Jeanne at all.

This absence is not for a shortage of documents. An astonishing archive exists online of all the documents relating to Pierre Poivre in Mauritius. It is a digital treasure trove. Every letter, every inventory, every report – transcribed, typed up, digitised and annotated by Jean-Paul Morel. There are thousands of documents here. The archive is searchable. The name Commerson prompts 114 documents. But in all the pages of letters, complaints, irritations, instructions and orders, Jeanne Barret's name never appears. I search for Baret, Baré, Barré, Bonnefoi, Bonnefoy, but there is nothing. I search for *femme*, *fille*, *valet*, *domestique*, *serviteur*, *gouvernante*. I can find no reference to Commerson's loyal and hardworking servant. She is invisible, like an item of clothing or a piece of luggage that follows him wherever he goes – unremarked and unnoticed – perhaps to be discarded when no longer required.

~

There are three large folio boxes in the archives of the Muséum National d'Histoire Naturelle, filled with images of the plants Commerson and Jeanne collected. They are drawn in exquisite and precise detail by Paul Philippe Sanguin de Jossigny, mostly pencil sketches, some in ink. Jossigny frames his images beautifully, sometimes cropping larger specimens so that more detail can be seen – giving the images a modern look.

Jossigny arrived in Mauritius as aide-de-camp to the new governor, François Julien du Dresnay, better known as Desroches, in June 1769. His artistic talent was immediately recognised, and Desroches allowed him to work for Commerson alongside 21-year-old Pierre Sonnerat, the great-nephew and godson of Pierre Poivre. Commerson was very happy with his two illustrators, 'who make me the most beautiful ichthyography that has yet appeared'. The feeling was not mutual. Jossigny found Commerson too exacting and demanding. It took all of Poivre's tact and persuasion to convince Jossigny to remain at his task.

As I search through the archive, it takes me a while to realise what these drawings are for. They are sorted by taxonomy: ferns, fungi, grasses, cycads and palms. Sometimes there are brief Latin descriptions on the back in Commerson's hand. A few also bear Commerson's signature – an ornate official signature, an imprimatur, complete with his doctoral qualifications. Or instructions for the engraver underneath the image – 'remove three leaves in shadow' or 'ensure all pores are illustrated'. Many of them are numbered as plates. Commerson was preparing the manuscript for a book – *Histoire Naturelle des Îles de France et de Bourbon*. These completed plates, signed off and numbered, show that he was well underway.

Now he had a team to help him with this grand task, perhaps he no longer needed Jeanne. But Commerson was single-minded in his drive to complete his life's work. He spent every penny he had, and some that he didn't, on this task, maintaining a driving pace that not everyone could keep up with. I cannot imagine that he would give up a patient, trained and hardworking assistant like Jeanne so easily. There was still plenty of work to do, drying and preparing specimens. If she was as valuable as everyone says, I can't imagine Commerson letting her go.

In the time Jeanne and Commerson were in Port Louis, Pierre Poivre's wife Françoise wrote to the incurably romantic and chronically depressed Bernardin de Saint-Pierre about his life choices. She advised him to retire to the country and live a simple life. Had she been a man of independent means, she said, she would not marry a well-off woman but rather 'the daughter of a labourer whose greatest merit would be to love me, to take care of my house, and to raise her children. Admit, sir, that this life would have its charms.'

Françoise Robin would have known Jeanne too. I can't help but wonder if Jeanne was the model for this loyal, faithful and hardworking peasant woman who might make someone's life so happy.

Amongst the Mauritian plants that Commerson found, one puzzled him greatly.

'It is a charming shrub,' he wrote to a friend knowledgeable about Mauritian botany. 'I am frantic, either because of the singularity of its leaves, or because it gives me a new kind whose character is unique. I am asking you to tell me, in regards to the leaves, whether the larger ones (the better formed ones) are at the top of the tree and whether the most irregular ones are set on the lower branches. The fruit, the fruit don't forget about it, please. In the past I named it *bonafidia* for this reason.'

Today this rare plant is not known by the name Commerson first gave it, but as *Turraea rutilans*, and is found only in the dense, high-altitude forests of the Mauritian mountains. It was once found on Réunion but hasn't been seen there recently. The leaves of the plant, like many in this genus, are 'heterophyllous', with many different shaped leaves. Early in his life, Commerson had been a devotee of a system of taxonomy based on leaf shape, rather than the Linnean focus on flowers and nuts. *Turraea rutilans* illustrates why the 'external' garb of the plant, or its leaf shape, can be a deceptive identification tool. Commerson named this plant after its perplexing 'vestitu' or 'foliis' – its clothing or leaves.

And it was for this trait that he named the plant *Baretia bonifidia*, after Jeanne.

Commerson found other plants belonging to this genus, all of which exhibited the characteristic variation in leaf shape. Jossigny illustrated three of them – *Baretia bonafidia* with its slightly rippled leaves, *Baretia eumonia* with oval-shaped leaves and *Baretia astata* with lobed or oak-shaped leaves.

The flowers of the plants in this genus are bisexual or hermaphrodites. I frequently find comments, on blogs and in serious papers, suggesting that these 'very doubtful sexual characteristics' were the reason Commerson named the plant after Jeanne. This seems like an unlikely thing for a botanist to have said. Around 90 per cent of all flowering plants are hermaphroditic, having both male and female organs. Commerson would surely not have found this trait either distinctive or even particularly interesting, for all his biographers do. I can find no record that Commerson ever said this.

There are thirteen specimens of this genus in the Muséum Nationale d'Histoire Naturelle herbarium collected by Commerson, representing seven different species: *Turraea rutilans, T. lanceolata, T. oppositifolia, T. ovata, T. thouarsiana, T. herbacea* and *T. lateriflorus*. None of them are recorded under the name *Baretia*. Commerson never published his description and no-one else published it for him, so the first published description is that of *Turraea*, with various incarnations as *Quivisia* or *Gilbertia*.

It is an apt plant to name after Jeanne – rare and difficult to find, with very little written about it, and a complex history of misidentification and reclassification. This history is literally written all over Commerson's specimens in the museum herbarium, affixed with multiple labels of differing names written by different hands. Many of Commerson's specimens only have a place of collection and his name written on them. Some don't even have that.

But *Turraea rutilans* has one small panel written in Latin, in ink that has bled brown into the paper over the centuries. All of Commerson's technical descriptions are in Latin, the common language of early science and botany, of which Commerson was a master. The writing is Commerson's, engraving the name of his hardworking helpmate, Jeanne, forever into the scientific record.

Baretia bonafidia – the true or authentic Barret.

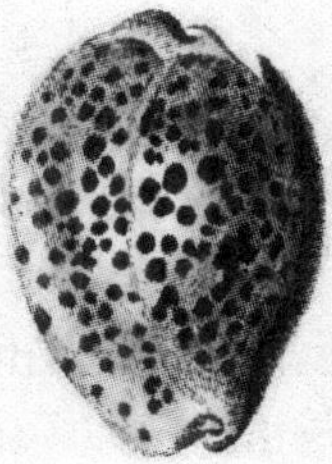

16

A WOMAN OF MEANS

Mauritius, 1770–1773

I AM SCANNING THROUGH the digitised Mauritius archives – page by page – when, quite by chance, a name catches my eye. The document is written in ornately obscure writing. It is formal and hard to read, but is clearly a land grant dated 12 August 1770. The letter is signed by Pierre Poivre and General Desroches, granting a concession for a block of land on the 'new street' that comes down the Petite Montagne, over which Fort Adelaide now towers.

The name on the document is not Commerson's. It is Jeanne Barret's.

> Under the authority granted by His Majesty and with His Majesty's pleasure, we have conceded to Dame Jeanne Baret

> a site located in Port Louis, district of Petite Montagne on which are two buildings, one in stone, the other in wood, which belong to her.

Almost nothing is known about Jeanne's life in Mauritius – where she lived, what she did, how she dressed or how she supported herself. But just as I had hoped when I started this project, there is new information to find in the archives, if we take the time to look. Here before me is previously unknown evidence that, just one year and eight months after her arrival here, Jeanne became a property owner in her own right.

It was not much – two small buildings – one of which had to be demolished so as not to obstruct the new road. The old French measurements translate to a property that was 5.8 metres deep and just 22.4 metres wide – an area, roughly, of between 117 and 131 square metres. Jeanne's property, I realise, was the same size as a single-fronted tiny three-room workers cottage I rented years ago in Melbourne.

Like most modern cities, Port Louis today is a mismatch of old and new, of varying architectural cultures and styles. Rampant development has battled against variable building regulations and controls, blocking drains and ignoring the flood-mitigation structures built in colonial times. In 2013, eleven people died in flash floods, which many blamed on poor development regulations.

It takes a while to see the layers of history along the modern streetscape. Old wooden French houses lean over street corners next to the upright pillars of English-era administrative buildings. Shanty shopfronts hang precariously over footpaths in makeshift expansions alongside traditional-style mosques and ornate Catholic churches. But the roads still follow old paths, the creeks and rivers still flow along their old courses beneath the ground, and occasionally surge forth swollen and angry through the streets and underpasses. Today, the 'new street' in Petite Montagne is named after Dr Joseph Riviere and leads from the citadel, past the peeling paint and packed windows of tiny shopfronts, towards the grand Jummah Mosque.

I find an old photograph of historic houses on a nearby street from the 1920s. The Citadel rises up in the background. They are old houses even then – their steep slate roofs, decorative eaves and finials and wooden shutters clearly marking their French architectural heritage, for all they are small, cheap and modest. Stone and timber patch together thin buildings just one or two rooms long. Shuttered gables make use of ceiling spaces, overhung by the spreading canopies of tropical trees that still manage to squeeze between the urban sprawl. This must have been the kind of house, the kind of streetscape, that Jeanne moved to and took her next step, away from her impoverished peasant childhood towards a more prosperous future, to financial security and respectability as a property owner.

My first thought on seeing this document, with Poivre's dramatic flowing signature inscribed beneath, was that perhaps Poivre had stepped in to tidy up the messy personal arrangements of his friend Commerson. But this is not a personal bequest from Poivre. This is one of dozens of standard formal documents that he signed, along with Desroches, as part of their everyday bureaucratic tasks in the colony. Why do I assume that it was Poivre or Commerson who arranged this land acquisition rather than Jeanne herself?

These land concessions were not purchased, but rather granted under conditions of use (agriculture, industry or residential) for twenty or fifty years. Only the income generated from the property was taxed.

Perhaps it was Jeanne herself who requested this property. She may well have realised that she could no longer maintain the required level of respectability while living with Commerson and acting as his servant. Some degree of separation must be observed for propriety. But whatever her motivation, one thing is clear – she was living in Mauritius as a woman, in her own home and with some degree of independence from Commerson.

It's a shame, I think, that mammals aren't hermaphrodites. It's such a common strategy in so many other taxonomic groups, particularly molluscs, and comes in so many shapes and sizes. I think we've missed out. Simultaneous, cyclic and sequential – some species are both male and female for their whole lives, others swap back and forth between the sexes, and many start out male and then, when they reach a certain size, become female. Females need to be bigger because they produce eggs, which are larger and more costly to produce than sperm. Female argonauts, for example, are eight times larger and 600 times heavier than the males, which cling to the females and leave a detachable arm to transfer sperm to the female, often mistaken for a worm or parasite. They have reduced themselves to an appendage.

Nobody really pays much attention to the life histories of shells, unless they are commercially important, so I'm not entirely sure which ones are sequential hermaphrodites or not. But I know which of my shells were old animals. You can tell by the weight and the heft. My favourite tiger cowrie must have reached a venerable age. Its shell is thick, smooth and glossy, like well-glazed pottery, quite different from the thin, slip-coated younger shells. I wonder if it was a female. Tiger cowries are supposed to be either male or female, no switching, but I wonder if anyone really knows. No-one seems to have done any ecological studies on them since the 1970s.

Maybe it's just one of those things we assume because no-one has looked carefully enough yet.

Within two months of Jeanne obtaining her house, Commerson was ready to leave for Madagascar. He had been planning to go for some time but he had suffered from several bouts of ill health – gout, rheumatism and dysentery – that kept him confined to his room. Madagascar was also dangerous to visit between December and May, when 'the season of epidemics' reigned.

'I embarked from Port-Louis, Isle of France, October 11, 1770,' wrote Commerson. 'I was accompanied by Mr. de Jossigny with the task of drawing everything I find most remarkable in the field. Mr. Poivre has left me nothing to be desired on leaving in the way of facilities and comforts for the trip.' His expedition was large and well funded.

There is no evidence, either on this trip or on the subsequent trip Commerson made to Réunion (then known as Île Bourbon), that Jeanne travelled with him. The timing of her land grant two months before his departure makes me think that they had planned to secure Jeanne's future on her own.

Whatever the rationale, it seems most likely that from here on, Jeanne's collecting days were done. Perhaps she was disappointed to be left behind. Or maybe she was too busy and occupied preparing for her own new adventures ahead. It could well be that she was simply happy to finally be in command of her own journey.

Commerson's plans are confusing, but it seems he left for Madagascar in October 1770 and arrived at Réunion in December of the same year. He stayed for a year before returning to Mauritius after a trip to Réunion's famed volcano in December 1771. All in all, he was away from Mauritius for little more than a year.

He made good use of his two months collecting on Madagascar. Even though he was restricted to the vicinity of Fort Dauphin (now Taolagnaro) in the south-east, the records of the global herbarium reveal 331 plants collected by Commerson in Madagascar and a further 496 from Réunion. His notebooks contain written descriptions of 68 Madagascan species. Commerson's contributions to the region's botany would have been substantial, had these descriptions ever been published. But this collection seems to have been completed without Jeanne's assistance.

Commerson had plenty of other assistants now, including a slave boy from Mozambique with an eye for plant collecting, despite Commerson's disapproval of slavery. Jossigny, however, soon had enough of working for the demanding botanist. This was not the only bad news. While Commerson was on Réunion, the authorities in Paris decided he should not continue his work, because another botanist had already been sent to Madagascar and it would be an unnecessary duplication of expenses. After extensive entreaties from the administrators of both Réunion and Mauritius, Commerson was finally approved to continue his work, but the stress took its toll.

When Commerson returned to Mauritius in January of 1772 he was exhausted but satisfied with his work.

'I do not know anything that I'm happier with than this job,' he wrote to Lalande. 'Nature has given Europe the weak examples of what she can do. It is at Réunion, as also in the Moluccas and the Philippine Islands, that Nature has fixed true pyrotechnical furnaces and laboratories. I have collected astonishing observations on their phenomena, of which the public may expect a large quarto volume, after I have given the Academy the first fruits of my labours.'

~

Jeanne's time on Mauritius coincided with a nexus of voyages through the Indian Ocean, of which she was one small part. Bougainville, Kerguelen and Marion du Fresne all passed through here. Mauritius was their staging post, a French safe haven from which to launch or recover from a daunting voyage into the alien, the exotic, the other and the foreign. This isolated outpost in the Indian Ocean sat at the crossroads of many journeys. I am accustomed to thinking of them as distinct, separated and linear, and yet here they all intersect, interfere with one another and go their separate ways.

There is a story that when Ahutoru was in Paris, he came across a mulberry tree. In Tahiti the paper mulberry is grown in every household plantation for making the *tapa* cloths that play such a central role in Tahitian ceremony and culture. Ahutoru wrapped his arms around this beloved tree and wept in memory of his homeland. It was time for him to go home.

He returned to Mauritius while Commerson was still away in Madagascar, entrusted to the good care of Pierre Poivre, who had befriended him when he first passed through on his way to France in 1768. Poivre was impressed by how Ahutoru remembered everyone who was kind to him, and how enthusiastically he sought out all his old friends. Surely that would have included Jeanne, given that Ahutoru, according to Vivez, 'took pleasure in being wrapped, powdered and dressed by her, which she did with good grace' and 'regretfully' parted from her when he transferred to the *Boudeuse*. In a town of some 15,000 people, 80 per cent of whom were slaves, surely over the course of twelve months they would have run into each other.

There is no record of what Ahutoru thought of his time in Paris, or Mauritius, or on the French ships. His Tahitian vocabulary was documented, but nobody recorded the stories he told to the great amusement of his friends. I am surprised that there are no portraits and few descriptions of him. Like

Jeanne, he is both invisible and unknowable – a tear in a carefully constructed historic painting that suggests an absence without telling what is missing. Few people even mention him.

His time in Paris had changed him. Bernardin de Saint-Pierre reported that on his way to Paris, Ahutoru had been 'open, gay and a little of the libertine', but the man he observed on his return, while still very intelligent, was reserved and polite. He had learnt the rhythms of Parisian society – not just the airs and dances he enjoyed, but also the measure of its time in the form of a watch which dictated the 'hour of rising, of eating, of going to the opera, of walking, etc.'

The astronomer Alexis-Marie de Rochon, who habitually used the title Abbé, felt that Ahutoru had absorbed the worst aspects of Parisian society.

'He had acquired to some extent the art of flattering the men he felt he needed. He studied them so as to provide clever caricatures,' reported Rochon. 'This Indian was of a frivolity which passed all measure . . . he said that in his country man was born to laugh and amuse himself.'

Everything I read about Ahutoru seems to fit the description of a member of the elite social class the Ariori – his dancing and chanting, his frivolity, the tattoos on his thighs and his beauty, his interest in sex. These are possibly not interests that the Abbé Rochon shared. I would be fascinated to know what Ahutoru thought of French society. Bernardin de Saint-Pierre must have wondered this too.

'Ahutoru seemed chagrined at his long stay in Mauritius. He walked, but always alone. I perceived him one day in a profound meditation, looking at a black slave at the door of the prison, round whose neck they were riveting a large chain . . .'

Ahutoru never returned to Tahiti and never saw the waters of his Pacific Ocean again. It was a year before Marion du Fresne agreed to take him home, via Madagascar, but he died en route of smallpox. News of this death would have travelled back to Mauritius quickly. I wonder if Jeanne was saddened to hear of Ahutoru's death, or sympathised with his inability to go home. He may well have been a confronting figure – a large and powerful man who expressed his sexual attraction to her in unambiguous terms. But he had also respectfully acknowledged her position as someone who had chosen to step outside the social confines of her gender. That, in itself, must have been a rare thing.

It was not just the arrival of people to Mauritius that unleashed a torrent of extinctions but the associated raft of introduced species, too. Clearing the land replaced the complex ecologies of forests, grasslands, mangroves and heaths with simplistic monocultures. Plagues of insect pests fell on these new crops, uncontrolled by the predators and constraints that had previously limited them. And like the old woman who swallowed a fly, the solution was always to introduce more new species.

'It is a misfortune that we do not have here more birds destroying the insects,' declared Commerson to his friend Joseph-François Charpentier de Cossigny. 'This island offers the spectacle of great forests without any woodpeckers. These are the great enemies of cariats [beetles], ants, small and large caterpillars. What service would we not give to this colony if we could to introduce species of shrikes, Dominicans, tyrants, flycatchers, woodpeckers, kingfishers, and other insect-eaters which never attack the grains; small hawks, butcherbirds, the birds of the night to balance the multiplication of granivorous chicks; as well as innocent snakes to destroy rats?'

It was a popular notion: Cossigny thought giant rats from Thailand might be a good idea to manage the plagues of grain-eating domestic rats. Or more cats to eat the rats. Sonnerat was keen on introducing more birds of prey and shrikes, even while the Mauritius kestrel (*Falco punctatus*) declined to the point where it was the rarest bird in the world. Commerson also wanted frogs, 'to purge the stinking waters of the prodigious quantity of larvae that swarm there'.

Fortunately many of the introduced species didn't take, despite the best efforts of the acclimatisers. But even so, Mauritius today still struggles to deal with the impact of cats, toads, monkeys, pigs, rats and deer, just as other island ecologies like Australia contend with their own plagues of cats, foxes, horses, camels, pigs and buffaloes.

I have a particular interest in feral animals, although it wasn't what I thought I'd work on when I was a child. By the time I left school, I had abandoned my dreams of being a marine biologist. I thought I couldn't do science because I was bad at maths and had not taken physics or chemistry. That wasn't true. I took politics and psychology instead, quickly discovered animal behaviour and, by the time I started postgraduate studies overseas, I found myself unexpectedly reclassified as a zoologist. I toyed with different projects – rats, hedgehogs, meerkats and South American bush dogs – but eventually settled on studying the impact of feral American mink on seabird colonies in the offshore islands of the Outer Hebrides. Studying feral animals, I imagined, would be a useful skill when I returned home to a country overrun with them.

I had no idea what the Outer Hebrides of Scotland would be like. My only expectations were shaped by the New Hebrides, the colonial name for Vanuatu, and a belief that the main impediment to field work was too much heat. Neither of these proved useful on the windswept North Sea islands that rarely reached above 20 degrees Celsius in summer. But as I sailed from island to island in a little gaff-rigged cutter kindly lent by a resident – measuring and monitoring tiny tern chicks as they grew; identifying fish bones in mink scats; watching otters dive for fish in the lochs; and eating fresh scallops from the Sound – I realised that I was not so far from what I had once imagined when I had dreamed of being a marine biologist, after all.

In early 1773, there was bad news for Commerson. His friend and protector, Pierre Poivre, and his family, were leaving the island and Commerson himself was seriously ill with dysentery.

The new intendant was Jacques Maillart du Mesle. What he may have lacked in charm or flamboyance, he made up for in solid and reliable administrative skills. He could not particularly see the need for a botanist and, more importantly, he could not find mention of one in his budget either.

Maillart demanded that Commerson move his collections out of the intendant's residence, forcing Commerson to purchase a new property at 194 Rue des Pamplemousses. It seems that Commerson did not have the ready cash to pay for this house and did so on a promissory note.

Maillart was not entirely cold-hearted. He instructed Jossigny to continue his work with Commerson for some months before allowing him to return to work in Réunion. But Commerson did not seem grateful for these leniencies.

'I found myself as another Prometheus,' he complained, 'nailed to the rock, and I must, with added misery, begin to move from the small apartment in the Intendant's house, that I have always occupied. Judge what it's like to move natural history collections for a man who cannot walk two hundred paces without being out of breath.'

Maillart lost his patience.

'He did not enjoy many friends nor public esteem,' Maillart told the minister. 'He was considered to be very debauched, and people regarded him as a wicked man, capable of the blackest ingratitude.'

The ingratitude is probably a fair accusation, but the debauchery and wickedness is unexpected. Perhaps Maillart is referring to Commerson's admiration for Tahitian free love. Or perhaps, more locally, he is referring to Jeanne and the scandal of living in sin with a servant.

Commerson always planned to go home. He dreamt of being feted in Paris, at least, for his achievements. I imagine Jeanne did too, that she wrote to her family and looked forward one day to being reunited with them.

But then everyone on Mauritius claimed they wanted to go home.

'Each man is discontented,' said Bernardin de Saint-Pierre. 'Each man wants to get a fortune – and to leave the place. To hear them talk one would think the island would be again uninhabited, every man declaring he will go away next year, and some of them have held this intention for thirty years past, yet remain to make the same declaration the year ensuing.'

But Commerson never returned to France, never received the acclaim he deserved, never saw the son he left behind. Lalande always said that he had had a sense of foreboding whenever Commerson had mentioned working on his martyrology of botanists.

'When he mentioned that work to me I foresaw,' Lalande claimed, 'even then, that he, himself, would, one day, be placed in this list of martyrs.'

Lalande described his friend 'in a retired country place, where he had neither emulation, nor even society to animate him, passing whole weeks, night and day, without interruption, sleep, or repose, in botanical studies, and arrangements.

He came often from his botanical excursions, in a piteous condition, bruised with falls from the rocks he had been climbing, torn with briars, emaciated with hunger and violent exercise, and with narrow escapes from torrents and precipices, where his life was exposed to the most imminent danger'.

He could have been describing Commerson's last years on Mauritius.

In early May of 1773, Commerson was in the mountains above Port Louis, in the town of Ville Bague, when he fell ill. He was planning to take a boat to the Grande Rivière on the east coast of the island to recuperate, before travelling to Grand Port, further south. But instead he went to the house of Jean Nicolas Bezac, a surgeon in the district of Flacq.

Commerson was said to have had a deposit on his chest and feared that he was going to die. He apparently had his will made by the priest at Flacq although no-one has found a record of it. He had a few notebooks of natural history plants in a leather wallet and some 300 piastres.

He died at 11.45 in the evening of 13 May.

Maillart dutifully informed the minister and made arrangements to ensure that his papers and effects were sealed and an inventory of them made.

One of Commerson's biographers tells the story of his death relayed to him by a member of Commerson's family by marriage, handed down over generations.

'He had a faithful servant left,' Montessus reports, 'who had witnessed all his troubles, all his dangers. His gentle hand was there to render him great service: his word was there for him to bring consolation and exhort him to hope. It is often enough, indeed, when abandoned, for a zealous servant to provide relief from human miseries and Jean Baret was this servant.'

It is a lovely story, a sweet imagining, of loyalty and servitude. And it is retold in various guises throughout the stories of Jeanne and Commerson.

But I'm not sure it's true. There is no indication that Jeanne was either travelling or living with Commerson any more. Bezac does not mention Commerson's servant travelling with him; in fact, he explicitly says that Commerson saw no-one but the priest.

Jeanne may not have been at Commerson's side when he died but there is no denying the constancy of her loyalty. Loyalty, virtue and industry are the three traits mentioned by every single first-hand account of Jeanne. Even if she was not with Commerson when he died, I think it is fair to believe she would have wanted to be.

Jeanne certainly knew when Commerson had died. Just four days later, she wrote to inform Clériade Vachier, their friend in Paris who was in charge of Commerson's apartment and was, in effect, the executor of his estate. The letter is mentioned by Vachier, briefly and in passing, in the legal documents about Commerson's estate. Did she write the letter herself or was it dictated to a letter writer? Did she write with the sad duty of informing a close friend of their mutual loss? Or was this a business letter to initiate the execution of Commerson's will and remind Vachier of her belongings in the apartment and what belonged to her? I am tantalised by the prospect of this letter, which offers the promise of finally reading Jeanne's own voice, her own words. Such letters reveal so much: conscientious care or a wavering scrawl, tight lines squeezed onto cheap paper or grandly spaced on quality cotton paper, stains, smudges, erasures, misspellings and even tears from a centuries old grief.

I doubt we will ever know. Perhaps the letter is still somewhere, in the Vachier family or departmental archives. But so far, Jeanne's letter has not been found.

~

It is impossible to say what the status of Jeanne and Commerson's relationship was at the time of his death. The true nature of their relationship, personal and professional, is lost in time. But for all their relationship is concealed, I am certain that Commerson retained a deep and abiding respect for Jeanne. He told us himself, in his own distinctive and idiosyncratic voice, of the true value she held for him. He even offered an account of the difficulties she faced as a woman, and very clearly declared that she successfully defended both her health and her virtue.

This note is written in Latin, in his notebooks, which have not been digitised, analysed, published or even fully transcribed. The note in the text that should have been published with the plant *Baretia bonafidia* records the words Commerson chose to put on the public record about his assistant and companion.

> This plant showing deceiving leaves or clothing is named for that heroic woman who changed into manly clothes and with the mind of a woman, traversed the whole globe, a thirst for knowledge as her cause, daring to cross land and sea with us unaware. She so often followed in my footsteps and those of the illustrious Prince Nassau crossing with agility the highest mountains of the Strait of Magellan and the deepest forests of the southern islands. Equal to the armed hunter Diana and the sage and severe Minerva, she evaded ambush by wild animals and humans, not without risk to her life and virtue, unharmed and sound, inspired by some divine power.

Fierce, athletic, determined, intelligent and honourable. Could any woman ask for a better epithet to be left for her in the scientific literature? Could any collector, collaborator or partner ask for a more sincere acknowledgement?

Commerson concluded his dedication with these words to his companion:

> She will be the first woman to have made the complete turn of the terrestrial globe, having travelled more than fifteen thousand leagues. We are indebted to her heroism for so many plants never before harvested, all the industrious drying, so many collections of insects and shells, that it would be prejudicial for me, as for that of any naturalist, not to render her the deepest homage in dedicating this flower to her.

17

UNEXPECTED TREASURES

Mauritius, 1773–1775

I wonder if Commerson mentioned Jeanne in that last will, in his final words in Flacq. I can find no record of the will and no-one else mentions it. I imagine it would have included further instructions on what was to happen to his collections on his death. But I also wonder if he repeated the request he made in his Paris will, for Jeanne to be given time 'to put in order the collections of natural history that are to be sent to the Royal Cabinet'. Did he mention her, leave her instructions or bequeath anything to her?

Whatever Commerson's intentions may have been, there is no evidence that Jeanne was involved in organising his collections. Jeanne had her own affairs to manage. She was busy running one of the many inns in the port quarter of Port Louis. She must have been doing well, for she held one of just six exclusive licences (at a cost of 500 livres) to hold billiards and serve tea and coffee as well as alcohol. The wine and brandy was imported from Europe, arak from India and the potent local rum made from sugar cane was known as *guildive*, from the English for 'kill devil'. In early December of 1773, the local paper reported that Jeanne was fined 50 livres for serving alcohol during Sunday mass, her conviction serving as a warning to the other bars in the area.

When Jeanne arrived in Mauritius she had been employed by the navy for two years and by Commerson for several more. Presumably, like every other crew member who left at Mauritius, she would have been paid the salary owed when she left the expedition. She may also have brought whatever small savings she had from France with her.

Like all sailors she would have used any opportunity she could to make money on the journey. Everything extra on the ship could be bartered for a price, even the slops from the kitchen or the rats caught in the bilge. Everyone was entitled to their slice or cut of the profits, as long as the captain was willing to turn a blind eye. The officers too had their own illicit opportunities. Commerson had complained bitterly about the goods brought on the *Étoile* to trade, even though he did exactly the same thing. Why would Jeanne have been any different?

Whatever she was doing, by the time of Commerson's death she was managing her own life, on her own terms.

In the Mauritian archives there are two handwritten inventories of Commerson's collections, the first of everything in his house in Rue des Pamplemousses in March 1773, and the second detailing the collections shipped back to France in November. There is no system to the packing. It is completely disorganised and disordered. Surely no biologist would pack a collection like this. Fortunately Jean-Paul Morel has transcribed and typed the list, otherwise I would struggle with the vague and cryptic descriptions.

There are 34 cases of material in all, in varying states, most of them smaller than a tea chest. They often contain 'herbaria' – pages of pressed and dried plants. Jeanne's careful work drying the specimens has stood the test of time.

The same cannot be said for the other specimens. The birds and fish in the inventory are described as 'debris', 'bits' or in a 'bad state'. Case 20 contained 'all the natural history drawings and a manuscript that we forgot to put with the others'. Case 16 contained 'a school bag in which there are dead fish and birds'. The inventory reveals an eclectic mix of crabs, coral, sponges, stony madrepores; samples of wood, seeds, roots, fruit, gum; stuffed and pickled fish, insects, reptiles, 'a penguin from the Southern lands', turtle scales; lava, pebbles and rock crystals. There are plants and animals, fish and fowl, rocks and miscellanea.

It does not sound like the largest and finest collection of the eighteenth century. Perhaps this is only a fraction of his work, or perhaps they were simply poorly packed and described. Sometimes it takes an expert eye to appreciate the treasures of nature. Even so, I am surprised there is no mention of skins or skeletons and only a few jars for preserving. The only stuffed specimens are some filefish, the penguin and, strangely, a '*cajou*', which translates as either a cashew tree or perhaps an owl.

Commerson had already lost some of the collections he sent back from South America to France. His death left his remaining collections even more vulnerable. There was no-one in charge to keep them safe, make sense of them or ensure that Jeanne and Commerson's legacy was secured.

~

I really want to know what happened to the shells. These inventories prove once and for all that Jeanne and Commerson did indeed collect shells. Commerson himself thanks Jeanne specifically for her collections of plants, insects and shells. They did exist, so where did they go?

There are four boxes in the collection listed as containing shells, two of them quite large. The inventory doesn't provide much detail. One box also contained madrepores or stony corals, another small box contains Madagascan shells, while Box 6 also held four herbaria without tags and a smaller box of large and small crabs.

One of the boxes contained '*un casque*'. It could have been a helmet or even the hard head shield of a cassowary. But when I read '*casque*' I think of *coquille d'un casque,* a common French name for helmet shells. And it brings to mind the giant helmet shells I saw piled high on a roadside market stall outside the Pamplemousses Gardens. The largest of the helmet shells, *Cassis cornuta*, the horned helmet, was named for its vague resemblance to a Spanish conquistador's helmet. These are the panzers of the mollusc world, heavy robust shells that patrol the seafloor on a thick muscular foot.

They hunt in slow motion – quite literally at snail's pace – smelling out prickly sea urchins and sea stars with two sightless antennae. Their hunting reminds me of naval sea battles in the age of sail – paralysingly slow and strategic, played out over hours or days, and with a growing sense of inescapable doom. Yet even predators have a kinder side. I once read an account by a diver who noticed a large upturned helmet shell unable to right itself. Slowly approaching in the distance were three other helmet shells. Curious to know what they would do, the diver decided to check on them later.

On his return, he was surprised to find the helmets had aggregated, but rather than attacking or eating their helpless associate, the arriving snails seemed to be helping it. They wedged themselves on either side of the stranded creature, apparently helping it to flip back into an upright position. The diver was astonished. Once upright, the shells dispersed, each trundling off on their separate ways.

It makes me wonder what else we would find out about the lives of seashells, if we took the time to look, if we could slow our hectic lives to their pace.

The cargo was shipped on the *Victoire*, which left Port Louis in Mauritius on 24 November 1773 and arrived in France three months later. Jossigny accompanied the collection home and handed everything over, including the drawings, to Buffon and Bernard de Jussieu.

'Jossigny, the draftsman who worked under M. de Commerson at the drawing of plants,' reported the captain of the ship, 'will be better able than anyone to give to France the information which might be needed.'

I doubt that. Surely Jeanne – a trained botanist and collector who had worked alongside Commerson, often as his sole assistant, for the previous ten years – might have been a better choice than an engineer and artist with no interest in nor affection for either Commerson or his work. She was there when many of the items were collected; she might have known what was collected when and where, which piece of paper belonged to which specimen, what Commerson might have meant by some of his cryptic annotations. Commerson did, after all, provide Jeanne with an additional year of salary, and occupancy of their home in Paris, to allow her time to organise his collections. Clearly he thought she was qualified for the job.

The inventory lists 34 cases, but Lalande's eulogy only mentioned 32 cases, 'which were deposited in the king's garden, under the care of Messrs. Jussieu, D'Aubenton, and Thouin, who are employed in examining and arranging them'.

Commerson's brother-in-law Beau suggested that these specimens, as well as the ones already in Commerson's Paris apartment, be assessed and any material suitable for the museum be retained; while the remaining specimens be sold for the benefit of Commerson's son Archambaud. He suggested that the plant drawings and descriptions be sent to Louis-Guillaume Lemonnier while the material on insect, birds, fish, shells and quadrupeds could be sent to Pierre Jean Claude Mauduyt, a medical practitioner with wide natural history interests. Both of these men could edit and publish the material.

The plant collections were reasonably self-contained. The herbaria were easy to keep together. The samples have their notes inscribed on the pages to which the plants are pressed. Commerson had already sent much of his collection back to France before he died.

Of the plants alone, it has been estimated that Commerson sent some 30,000 specimens back to France. At the time this was said to be the largest collection ever acquired by a single individual and included as many as 5,000 species, of which perhaps 3,000 were entirely new to science.

Not all of the material arrived at its destination, and not all of it was as new as Commerson had hoped. Antoine Laurent de Jussieu wrote to Archambaud in later life, clarifying that Commerson's collections contained many duplicates. In total there were only about 4000 different species, and of these two-thirds were already known. He considered that about 1000 might be new to science.

'However, we must observe,' he added, 'that we do not yet know 20,000 plants and that a botanist who added a twentieth of his contemporaries should be put in the number of those who occupy the first ranks of science.'

It was Jussieu who published many of Commerson's descriptions, in *Genera Plantarum* (1789). He published 37 of Commerson's genera with appropriate attribution – for example, listing *Bougainvillea* Comm. ex Juss. – and he published a further fifteen new genera based on Commerson's material. In the herbaria of the Muséum National d'Histoire Naturelle in Paris today, there are more than 5000 individual specimens designated as having been collected by Commerson, and of these 73 are holotypes – the original specimen from which a new species was described. According to the International Plant Names database, 206 plants have been published under Commerson's name, including both genera and species. And even though not all of these names remain valid today, it is a legacy that does, indeed, place Commerson in the first rank of his botanical peers.

'It is astounding,' the great palaeontologist Georges Cuvier later reflected, 'that one man should have been able to do so much in so short a time and in a tropical climate.'

Perhaps that was because these collections were not the work of just one man, but a man supported by a remarkably hard-working woman. Much of Commerson's success must also be attributed to Jeanne.

The other collections, however, have fared far less well. Commerson has been recognised for his work on fish, reptiles and whales. Once Lacépède located Commerson's notes after Buffon's death in 1788 (some say they were in Buffon's attic), he published many of the fish descriptions. But I don't think many of the bird or mammal specimens have survived. The stuffed birds were lost during the revolutionary years, not by rebellion (for the rebels too valued science) but by the over-enthusiastic application of sulfur – a preservation technique recommended by Mauduyt that destroyed the specimens they sought to protect. Buffon did include Commerson's descriptions of birds, and some notes on mammals, in his publications, but for some reason these new descriptions of species were not adopted.

Of all the other specimens, including the shells, I can find no trace, no connection with Commerson, let alone Jeanne, whatsoever.

The entrance to the laboratory at the Jardin des Plantes in Paris is from a side street that runs along the outer wall. An assortment of buildings has accumulated over five centuries, connected and linked, leaning together in mutual support to form some kind of cohesive whole. We pass through an arch built large enough to allow the entry of loaded carts and horses and my guide directs me left into a heavy side door that opens into a well-worn institutional staircase about as welcoming as an abandoned mental asylum. It does not take many floors, corridors and offices before I am hopelessly lost and realise I would never find my own way back without my escort. I try not to think about fire exits.

The malacology labs are at the top of the building, crowded into an attic space with sloping ceilings. There are no gleaming white benches, no chrome fittings, no white coats or hi-tech equipment. There are creaking wooden floors with patches of worn linoleum, rows of mismatched desks united only by their piles of papers, books and reports. The walls are lined with cupboards, ranging from serviceable grey aluminium to the occasional ornate mahogany cabinet fine enough to grace a grand historic library. They are filled with narrow drawers that pull out to reveal boxes of shells: old, new, colourful and faded.

I love the backrooms of old museums and their chaotic rabbit warrens. My first full-time job, after completing my doctorate and returning to Australia, was with the Museum of Victoria, as it was then known. A dream job, really. I was destined, it seemed, to be a writer rather than a field worker, after all. I was employed as an 'essayist' – to write stories about the museum's natural history collections to help in exhibition development. The vast unseen collections of that museum, just like those of the Muséum National d'Histoire Naturelle in the Jardin des Plantes, are a treasure trove of such stories – historical, biological, cultural and personal. Too much for just an essay or two, so I wrote a bookful instead. It was one of these stories that first led me to the Paris museum – in search of a trigonia shell collected by François Péron on the Baudin expedition, and specimens of the same shells collected live by Jean-René Quoy and Joseph Gaimard on one of Dumont d'Urville's expeditions.

Despite their antiquity, these garret offices at the Muséum National d'Histoire Naturelle resemble those of curators the world over: crowded with strange jars and equipment piled on unsteady stacks of paper and with never-ending need for more storage. We move past cupboards that serve to break up the open spaces into smaller rooms, and I am introduced to various researchers as we pass, the usual multicultural mix of modern science. In broken English and French, we converse about their research topics and seek out common ground and interests in our work.

Eventually, I am taken to the furthest corner of the labs. The cabinets are nearly all old here, a motley and beautiful assortment of carpentry skills with carved doors and hidden drawers, ornate and curvaceous. Most were custom-built for the collections they have held secure since the seventeenth, eighteenth and nineteenth centuries, form made beautiful by function and craftsmanship. Here I find the trigonias collected by Quoy and Gaimard, suspended beneath tiny glass bouys in a jar. And the original broken bivalve found by Péron on Maria Island in Tasmania on the Baudin voyage, its name written out in the hand of Jean-Baptiste Lamarck himself. But I will not find any shells from the Bougainville voyage in this collection. The curators are adamant that they are not here.

And yet, if they did not end up here, where did they go?

I saw a copy of Jeanne Monnier's biography of Commerson in a maritime museum bookshop while I was in Brest. I had gone to Brest to visit the archvies, but they were closed for a saint's day that I hadn't anticipated so I spent my time visiting the maritime museum in the fortified chataeu, looking at models of ships and visiting the Lapérouse memorial. I was tempted to buy the Monnier book, until I remembered that I was already having trouble closing my suitcase and so I resisted, a decision I swiftly regretted. Now I can't find a copy anywhere in Australia and it is the most recent book to discuss Jeanne. In desperation, I order a copy online, which arrives in a week or two from a secondhand bookshop just down the road from the Brest maritime museum.

The book is a treasure trove of images and extracts from the archives. It is here that I find mention of a new document that I haven't come across before – a handwritten version of Commerson's bequest to Jeanne, which shows that in French (unlike Latin) he spelt her last name Barret too. I notice that a section in the book dealing with the research collections includes not only the plants, but also the fish, birds, insects and geology, concluding with a section labelled 'conchyologie'. It is the first time I have seen anyone discuss the shells collected by Jeanne and Commerson.

The section is short but references an inventory of items found when Commerson's Paris apartment was opened in July 1774 by Vachier for a panel of museum experts to inspect six cases and a barrel of natural history objects. Most of these cases contained shells, including many from the Magellan Straits, Rio de la Plata and Tahiti.

I am heartened by this evidence that the shells did in fact make it back to France. But I am still no closer to working out where they are. Of the museum experts overseeing this removal, most were botanists led by Antoine Laurent de Jussieu. Only Mauduyt had more generalist interests, mostly in electricity, but also in natural history as evidenced by his renowned cabinet of curiosities.

I can find nothing further to connect any of them to shell collections.

I have one box of images left to search among Commerson's archives. I have been through the plants – the grasses, palms, shrubs and herbs. And I have been through the birds, reptiles, fish, crustaceans, mammals and insects. I am slowly getting a feel for the structure of the images.

I am beginning to realise that Commerson was not boasting when he described the huge diversity of material that they had collected, and begged to differ that the world's biodiversity had even begun to be counted. He argued that there must be far more than the claimed 20,000 plant species in the world given that he had personally collected some 25,000 species, yet he knew there were some five or six times as many still to be found.

And not just plants. The drawings include Madagascan animals that are not always easy to come by – tenrecs, civets, genets, mongooses and even a fossa. There are far more birds than I expected. Some of them are carefully drawn, detailed pencil sketches – a gull from Rio de Janeiro, a toucan or a pheasant from undisclosed lands. Others are watercolours, often quite crudely painted. I assume these are colour plans, purely to inform the printer where to place the different colours on the engraving.

The drawings are for natural history atlases, well on the way to being ready for publication. The fact that so few of them were ever published is heartbreaking.

As I work through this final folder, there is still one major taxonomic group that I have yet to find. Surely here I will find the molluscs.

But when I open the folder, I only find more birds. The molluscs are missing.

There is only one shell drawing in all of the archives that I can find. It is, in fact, a hermit crab – *bernard l'hermite* – that has housed itself in a delicate harp shell, *Harpa articularis*. I have a particular fondness for harp shells. They are so fine and frail that they are hard to keep intact. They don't survive the rough and tumble of the surf, nor life in a box on a boat. The one in my collection was abandoned on a jetty pylon. I don't know who left it there but I have treasured it ever since.

I'm not sure a harp shell is the best home for a hermit crab. I'd prefer something sturdier, I think, but the delicately patterned ticking of the shell suits the russet hairy legs of the crab, with its curious bulbous eyes. It's a rust-red sanguine pencil sketch – lively and quick – not one that seems intended for publication.

It is possible, I suppose, that Commerson did not consider shells worth including in his publications. Many shell collections at the time included plates arranged in decorative patterns, suggestive of the strange fashion for shell-encrusted miniature churches, windmills and houses that seemed to take pride of place in many of the seaside towns we visited in my childhood. Or the glued-together shell animals, ladies or lamps made out of bottles and baler shells.

Commerson was well aware of the value of shells as 'objects of curiosity and luxury, spread out in the offices of wealthy amateurs, who want to conceal their ignorance by presenting them delivered modified, polished, coloured and altered in a thousand ways'. Such collections negated the true value of the specimens. Maybe the shell craze was enough to put a serious naturalist off them completely, for fear he might be taken for an amateur. But Commerson was unlikely to suffer from such concerns.

'They are never more truly valuable to naturalists than when they are covered with their marine cloth: this is how they must be harvested and preserved,' he wrote in his guidelines for naturalists.

There are still so many new species to be found, among the molluscs as well as other marine invertebrates, that a museum curator once told me that if you collected any handful of sand from the shore, you would be almost certain to find an undiscovered species in it. New species of molluscs are still being found and described today. Lamarck and Jean-Guillaume Bruguière and many others continued their work on new mollusc species through the eighteenth century. Quoy and Gaimard brought back new species from Dumont d'Urville's voyage. Baudin caustically commented on Péron's collections of broken shells, which included the first extant specimen of a trigonia, from Tasmania, a family otherwise only known from the fossil record of the Pleistocene. Surely Commerson would have found new species among those he and Jeanne collected.

I know there were shells in this collection and I suspect there were drawings too. I know Jeanne collected them, I know they were shipped back to France and I know they made it to their Paris apartment. I would have expected them to have been given to Lamarck to work on, but there are no records of any specimens in his acquisitions register.

I suspect the shells were simply too popular for their own good. Bernardin de Saint-Pierre gave his shells to Madame Poivre before he left Mauritius, which she accepted with some reservation to put in her cabinet 'if ever the taste of having one takes me'. By the end of the 1700s, there were more than 600 shell merchants plying their wares in Paris. I can only assume that the shells were part of the collection that was sold for the benefit of Archambaud. The shells must have ended up in private hands, in someone else's collection, and perhaps any drawings went with them.

I cannot give up on the shells. From time to time I procrastinate by entering random keywords into my search engine – 'commerson', 'mollusc', 'coquille', 'shells' – hoping that some new combination of names, words or languages will locate an unexpected source. I search the catalogues of different libraries and digital archives – old books, science databases, newspapers and museums, flicking between languages. It's a fishing expedition. Mostly I catch nothing, old boots or a bit of weed, but occasionally there's a tug and a bite, and my attention is caught.

This time, I have put 'bougainville' into the equation and my search returns something interesting. It's a German paper published in 1872 in a *malakozoologie* or mollusc biology journal. 'Conchylien von Bougainville's Reise' – shells from Bougainville's voyage. The paper identifies shells that came from Bougainville's voyages in the collections of a wealthy Parisian collector.

Antoine-Joseph Dezallier d'Argenville was an influential man in Paris, secretary to the king and an advocate for the parliament that led to the Revolution. He was interested in gardens but also in shells. His office was renowned not only for his collection of artworks, but also for the shells that he believed should be the subject of a new science of conchology. His exquisitely illustrated and produced quarto books on the subject, first published in 1742, were much sought after, and new editions appeared regularly. The third edition, twice the size of the first, was published in 1780.

This book contains several plates of southern shells, and features shells from Tahiti which can only have come from Bougainville's expedition.

'The coasts of the island of Tahiti or Cythère have recently enriched cabinets of the curious and amateurs, of various shells,' he wrote.

These shells, of course, could have come from any one of many collectors on the voyage. Bougainville himself collected shells, as did many of the sailors. Dezallier d'Argenville mentions Bougainville's narrative, and the hammer shell they found in New Britain. But the list of other species is long and diverse. There are limpets, chitins, murexes, volutes and clams from the Magellan Straits. From Tahiti there are turban shells, oysters and scallops. There are pages of cone shells, dramatic spidershells and gleaming cowries.

And there is also a mention, from Tahiti, of '*le grand Casque de l'espèce triangulaire*' – a giant helmet shell, illustrated on the plates just next to a page of the most delicate array of harp shells.

Perhaps these are the shells Jeanne collected. Perhaps they are not. I don't know where d'Argenville's collections ended up, but I think the beautifully illustrated plates in his book are probably as close to them as I am going to get.

'Listen to the sea,' someone told me as a child, holding a shell to my ear.

But it did not sound like the sea to me. All I could hear was the sound of my own heartbeat, pounding in my head.

It has never occurred to me before that the molluscs are entirely silent. Such a large, ancient and highly diverse phylum and yet, of the 100,000 species currently living today, not one of them generates a deliberate sound. There is no reason why not. Sound travels well under water, from the ear-cracking pop of the pistol prawns to the subsonic crooning of humpback whales. The terrestrial land snails are similarly silent. Nor is it to do with lack of equipment. Molluscs have hard parts they could use to make sound, just as insects often do to call to mates, to warn about predators, to locate prey and to promote social bonds.

Perhaps it is just too risky for such generally slow-moving animals to draw attention to themselves unnecessarily. Perhaps the costs would outweigh the benefits. Few shellfish are fast enough to escape from a fish, or aggressive enough to fight back. And those molluscs that are fast and literally well armed, like the cephalopods, lack the hard parts required to make sound.

And so the molluscs remain silent, leaving us to follow only their faint trails in the sand.

I am not sure if Jeanne would have heard from her family at home while she was in Mauritius. Letters could have been written and read, if not by Jeanne and her sister, then by village writers who provided this service for a fee. Ships have delivered ad hoc mail services for as long as they have sailed, and formal mail services were established in Mauritius in 1772.

I wonder, then, if she knew about the death of her nephew Léonard, her sister's son, on 9 May 1773 at just ten years old. Did she think back to the birth of her nephew, the year before she fell pregnant with her own son in Toulon-sur-Arroux? She had seen so much in the last ten years, but she had missed so much as well. She had lost not just her own son, but the nephew she hardly even knew. Her sister only had Romain, her son from her first marriage – now a young man of eighteen – and her daughter Françoise who was just twelve years old. It cannot have been easy to be so far from her family when they needed her most.

A year later, Jeanne had happier news to share with her sister. On 17 May 1774, at the age of 33, she married Jean Dubernat, a 37-year-old drum-major in the Royal-Comtois who had served in the army for the last twenty years. Dubernat had been in Mauritius since 8 April 1770. I have no idea how he met Jeanne, perhaps at her work if she was indeed running a tavern. Port Louis was a small place, so they may have known each other for some time.

I have been trying to find a map of Port Louis in the 1770s. I hope this might tell me where she and Dubernat lived and where they were married. The online archives in Mauritius unexpectedly go offline. Eventually I discover that their portal is being renovated, that they will be offline for several months. If I want to consult the archives in the meantime, I am told, I am welcome to visit their office in Port Louis.

A friend puts me in touch with the local historical society, and Jean-Marie Huron kindly offers to find the maps for me. Along the way, he checks the notarial records for any other documents regarding Jeanne Barret. Unexpectedly, he finds not just the marriage certificate, but also the marriage contract between Jeanne and her new husband.

In the years since she arrived in Mauritius, Jeanne had acquired a small fortune. Dubernat brought furniture and clothes valued at 1000 livres and gave 4000 livres to his wife. For her part, Jeanne brought not only a house, slaves, furniture, clothes and jewellery, but 19,500 livres in cash. She gave a third of this to the 'community of marriage'. Two-thirds she kept under her own control.

Jeanne's fortune was a sizeable sum for the daughter of a labourer to have accrued on a servant's salary of just 12 livres per month. Even running a small tavern would have been unlikely to generate this small fortune. Most wealth in Port Louis was agricultural, but her property was too small for this. It's still possible she had land elsewhere and was running a farm, but there are other possibilities too.

I remember the illicit goods for trade on the ships, in which Commerson both participated and complained about. Jeanne had excellent connections among the trading vessels coming into Port Louis, not least among the Breton officers she sailed with. La Giraudais, the captain who, in all probability, had accepted and protected her on his ship, did not return to France with the rest of the expedition but remained in Mauritius. His wife and children soon joined him. Their last two daughters were born there.

I don't know if Jeanne was involved in trade, but if she was she seems to have enjoyed greater returns on her investments than La Giraudais, who sailed back to South America but lost his own and other's money in the process, although he was not blamed for the loss. The risks were high, but presumably the rewards were great. Jeanne's indisputable experience as an expeditioner would have garnered the admiration, grudging or not, of any captain coming into port. For anyone wishing to trade in goods in an isolated seaport, these were valuable connections indeed.

It's unlikely that she received anything from Commerson's estate in Mauritius. Maillart seized all of Commerson property as, in effect, the property of the state, using any spare funds to pay off Commerson's creditors and sending the collections to France.

But perhaps Maillart didn't seize everything. There is no mention of the diamonds that Commerson obtained in South

America and spoke of in a letter to his brother-in-law. They never appeared in the inventory or accounts of his estate. Commerson clearly considered them his insurance. We should not underestimate their potential value. After all, the English Pitt dynasty was said to have been entirely funded from the sale of one stolen Indian diamond. Who knows how much Commerson's diamonds were worth, what happened to them and if he left them in Jeanne's safekeeping. Or perhaps she had invested in her own. It's possible. Perhaps this was the source of the jewellery mentioned in her marriage contract.

This is no longer the story of a poor girl thrown off the ship at Mauritius, abandoned by her master, scraping a living selling bootleg liquor and running foul of the law, to finally be rescued through marriage to a soldier in order to limp back to France.

This is a rags-to-riches story – of a woman who not only sailed the world, but made her fortune and returned to France successful and independently wealthy. When she married Dubernat in the Saint-Louis chapel in Royal Street, alongside the tomb of Madame de Labourdonnais, it was Jeanne who was the catch. Their marriage was witnessed by five men, tradesmen and merchants of varying ages, a mix of young and old French arrivals in Mauritius. Not servants or labourers, neither upper nor lower class, neither wealthy nor poor: these were men more or less in charge of their own livelihoods and destinies. Jeanne had joined the middle classes – the petite bourgeoisie.

It was time to return to France. She later reported that they had to stay on Mauritius for six months after they married. I am not sure why. Shortly after they married, Dubernat obtained a concession for land, presumably a larger property where they might live together. It was a requirement of Mauritian land concessions that the owner develop it within three years or the property reverted back to the government.

But whether Jeanne had acquired her considerable wealth through property, trade, innkeeping or something else, over the nine years since she arrived on the island and since Commerson's death, it seems that she had done herself proud.

There is another surprise hidden in the marriage contract. Jeanne specified that a large sum of money – some 6000 livres – were to be sent to France and deposited for Aimé Eugène Prosper Bonnefoy, who was born at the Hôtel-Dieu, the public maternity hospital in Paris, in May 1766. So there was a child named Bonnefoy, after all, presumably a second child of Jeanne and Commerson, a younger brother of Jean-Pierre. Not a child born from the expedition in Mauritius, but in Paris in the very year they left.

I search the admission records for the Hôpital des Enfants-Trouvé for 1766, but there is no Bonnefoy listed. Whoever this child was left with, he was not on public assistance. With the loss of most of the records in Paris in the 1871 Commune, there are no parish records I can search for details of his birth, no declarations of pregnancy. I can find nothing on any of the genealogical pages.

The records of this child too, like those of his older sibling, seem to be lost.

But perhaps they were not lost so much as deliberately concealed? Among the legal papers of Commerson's estate, there is a statement by Vachier about extra payments made by him on Commerson's behalf to an undisclosed person. It seems like a lot of money – over 4,000 livres – and Commerson asked Vachier not to tell his family about it. There is no record of who the person is, or what the payment was for, but at some point the payments stop because the mystery recipient is dead.

Included in the list of expenses Vachier pays on behalf of Commerson is a death certificate for an *enfant protégé* – a child under the protection of Commerson. I am none the wiser as to whether this was the child of Jeanne and Commerson, but it seems increasingly likely.

As I search all the possible variants of his name, I wonder if Aimé was his first name, or if the sentence could be read as 'beloved' Eugène Prosper Bonnefoy instead.

From the language of the marriage contract it seems that Jeanne was uncertain if her son was still alive. If so, she was determined to leave him financially secure. The contract stipulates that even if Jeanne died before they returned to France, Dubernat was to uphold this clause for her. But for all her concern, despite the large bequest that should have secured his future, I can find no further records. Aimé Bonnefoy has left no trace.

After so many centuries, the archives, and Jeanne, continue to surprise me.

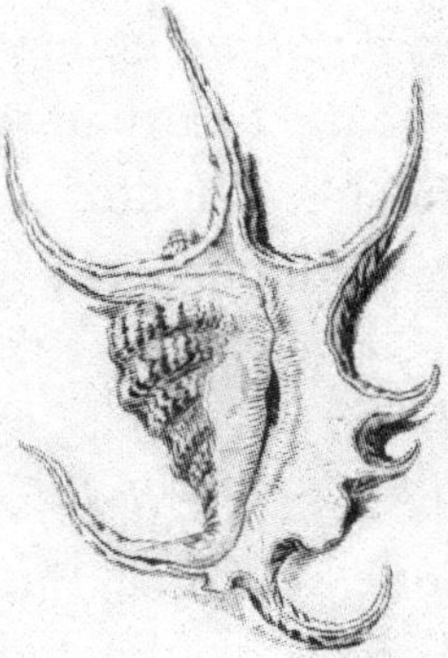

18

HOMECOMING

The Dordogne, 1775–1807

IT IS HOT ON the day we visit Jeanne Barret's grave. We can't stay long. Christele and I have a tight schedule. It has been several hours' drive from La Souterraine and we still have to get to Rochefort, 200 kilometres away, before nightfall. The countryside surrounding the township of Saint-Aulaye in the commune of Saint-Antoine-de-Breuilh of the Dordogne is unprepossessing – flat monotonous farmland broken only by motorways and powerlines. Small towns of close-shuttered houses resist prying eyes and deter visitors. We head for a cemetery at the church near the Dordogne River where Jeanne Barret is buried.

There is only so much that you can tell from an unmarked gravestone. You can tell if someone was wealthy enough to afford a fancy stone, or important enough for others to care for it. Not all explorers have the luxury of such memorials. Many lie in unmarked graves, like Baudin and Commerson somewhere in Mauritius, or Flinders who was unceremoniously paved over to make way for Euston train station, and only recently rediscovered and reinterred. Some never made it home at all, like Cook, du Fresne, Lapérouse and d'Entrecasteaux.

Jeanne Barret has done well for an impoverished peasant girl from Burgundy. She is remembered and commemorated, not just in the place she was born, but also on the other side of France in the place she died, where she lived out the last of her days as a prosperous landowner and merchant, in a large and comfortable home, surrounded by family and respected by friends.

I'm not really sure what I expect to find at her gravesite, or in the towns where she spent the last 30 years of her life. But I hope at least to find out how she came to live here, as a respectable and well-heeled citizen, in this small corner of France. And to gain some clue as to who she was, beyond the grand adventure of her youth.

Jeanne returned to France, but for many years no-one was quite sure when or how, or even exactly where. Perhaps somewhere in the Dordogne? I asked one of her biographers about her later years, but he admited his interest was mainly the circumnavigation and he didn't know much about her life on Mauritius or after her return to France.

The voyage was one and a half years long. A blink of an eye in a 60-year lifespan. Surely the before and the after is also important. I have been wanting to find out what it was about Jeanne that sent her on this voyage. And now I want to know what impact it had on her life afterwards.

Voyages can be influential, no matter how brief. I spent my childhood, for as long as I can remember, preparing to live on a boat, watching our boat being built, living on the half-built construction in the middle of the country, on the slipway, in the harbour at Port Lincoln. I spent barely three years actually at sea – and most of that in lengthy stays in a handful of harbours. We circled less than a third of the continent, from the middle of the south coast to the edge of far north Queensland, and then later to Papua New Guinea. Weather permitting, you could make that trip in a couple of weeks.

And yet, it is to this voyage that I return over and over in my own writing, in my reflections. The memories of our early years, eventful or ordinary, are preternaturally vivid, laid down in rich detail on a blank page, unlike the crowded busy notes of later life, scrawled illegibly in the margins of our mental journals. That short time I spent at sea has shaped my habits and tastes. I still stubbornly resist the constraints of the clock, repeatedly rejecting regular employment for the life of an independent but impoverished freelancer, tripping over the inexplicable regulations of public holidays, shopping hours and daylight savings, preferring to follow the rhythm of the seasons and the weather. I spend windy nights sleepless and disturbed, listening for movements in a house that withstands gales with barely a creak or groan.

I am waiting for it to shift on its moorings, for stone to release from concrete foundations, to drift free across the baffled surface of a distant ocean.

It has taken me years to stop myself from leaping to my feet, heart pounding, hands itching to release the cold metal of brake on the anchor chain, sending fathoms of iron into the muddy seafloor. My parents worry that this is some kind of trauma, but it's more like a childhood imprinting that might go unremarked if I still lived the life I was raised for.

Jeanne was only 35 when she returned to France. A youthful voyage, however remarkable and eventful, is hardly the end of a life. Take 53-year-old Susan Sibbald, for example, who left her ailing husband in Scotland to visit her sons in Canada in 1835, bought 600 acres and founded the community of St George in the Ontario wilderness. Or Edith Coleman, who began her career as a nature writer at 49 before rising to international acclaim. Women are often late bloomers.

Jeanne returned from her voyage to Les Graves just north of Saint-Aulaye in the Dordogne region. It was the home town of her husband, Jean Dubernat. I don't have any more specific directions or locations than the name on the map and I'm not confident of finding anything here. But we have driven half a day across France to get here so we have to try.

The roads are so small we keep missing the turnoffs, mistaking them for driveways or footpaths barely large enough to take a bike, let alone a car. But, weaving and reversing, we finally find ourselves in a small patch of countryside between two busy roads that both lead anywhere but here.

A cluster of houses crowds between the open fields. Not a village, barely even a *lieu-dit*. This is Les Graves. It feels like part of a farm, a collection of houses and outbuildings that has, over time, been converted mainly into accommodation. On the edge of the cluster are some older buildings, down a short dead-end road that feels uncomfortably like driving into someone's yard. But no-one shouts to see us off as we stop at the end of the lane.

It's more than the assortment of buildings, barns, houses and outhouses that makes it feel like a farm. There are no fences between the buildings and the farmland beyond, no flower gardens, no obvious sign of domesticity or suburbia. Ivy grows over the walls of the low barns, their few shutters tightly closed. An apricot tree in the middle of the lawn is laden with orange-gold fruit. A handful of plants that might be ornamental, or might just be weeds, outcompete the grass that tangles around their base. There are few odd pots and timbers stacked to one side. It's a tidy yard, but not one cultivated for appearances. This is a working property.

A single rose rises from a crack in the narrow concrete path at the base of an old house that is clearly under repair. Less decorative than simply indefatigable, as roses often are. The roses in my own garden are all but indestructible too, for all the frail and fragrant beauty of their luscious blooms. It reminds me of Diderot's 'strong soul in a frail machine'.

The French doors are open, the barnyard shutters downstairs are either ajar or askew, unhinged from the crumbling lintels and window frames that are being repaired and repointed. But the single shutters on the tiny upper-storey window just beneath the plain tiled roof remain firmly closed. I'm guessing an owner-builder is working here but I lack the courage to go and knock on the door. I have no way of knowing if this building has anything to do with Jeanne Barret at all, if anyone even remembers her, and it's just too complicated to try to find out.

We quickly take some photos, back out, and continue on our way, unsure of whether we have found anything or not.

I had searched for shipping lists from Mauritius and France. I expected them to be easy to find. After all, shipping lists to and from Australia are well documented – one of my Jaunay cousins has compiled and maintained a large online database of the ones for my home state. But as I search the lists, I realise that nearly all the records have just four countries as their destination – the United States, Canada, Australia and New Zealand. Shipping lists, it seems, are the particular obsession of the English-speaking colonies. For everywhere else the search is manual – through papers and microfiche, ship by ship, port by port, year by year. Not for the first time, I am reminded of my colonial heritage.

'Sieur Dubernat with his wife and his brother' arrived in Bordeaux on 26 August 1775, via the colony of Saint-Domingue (now part of Haiti). If this is Jeanne, her husband and his brother François, they were three of only ten passengers on the merchant vessel *La Sympathie*.

If Jeanne and Dubernat left the island six months after their marriage, they would have left in November 1774. But they didn't arrive in France until August 1775. It doesn't usually take nine months to sail from Mauritius to Europe. What were they doing in Saint-Domingue?

It could simply be because of the weather, or because it was the only vessel they could find passage on, but it is more likely to have been for trade. There was still a significant trade between Mauritius, the Caribbean and France in 1775. I worry briefly that they might have been on a slave ship, that perhaps they were involved in this lucrative trade. It was one thing to be allocated slaves in Mauritius, to which every landholder was entitled, but quite another to actively trade in them. Le Morne Mountain in Mauritius, with its dark compelling history of the slave rebellion and tragedy, rises in my memory. I remember, too, the East India Company Museum in Lorient, Brittany, with endless rooms dedicated to the treasures and history of this vast company, and only one small display acknowledging the slave ships on which the vast bulk of its wealth was amassed.

But there is no record of a ship called the *Sympathie* in the extensive transatlantic slave trade database of voyages. It is unlikely that Jeanne and her husband traded in slaves – more likely that their business dealt with the more pragmatic needs of everyday men and women: food, drink and clothes.

Entering the Gironde estuary and docking at Bordeaux, they would have returned to France close to where Jeanne's journey had started, nine years earlier, just north in Rochefort. The return was much warmer than the departure. The summer had been uncharacteristically hot and dry – enough to make the sheep pant, the puddles parch and the mud crack into shards. But surely not hot enough to trouble those from tropical climes.

The hollyhocks – fellow travellers from eastern lands – would have welcomed them home, sprouting in bright vigorous profusion from every roadside crevice. As Jeanne travelled inland, towards the hometown of the Dubernat brothers, the fields were filled with the crops, trees, birds and animals familiar to her from her childhood. It would have smelt like home, for the first time in years.

Within two months, in October 1775, Dubernat had bought a stone and timber-framed house in Sainte-Foy-la-Grande, a busy little town with narrow crowded streets. It was not a large house – two rooms on the ground floor above a cellar and two rooms on the upper floor with an attic under the tiled roof. It looked west, onto the garden with a well that they shared with their neighbour. The house was located in a retail area, on what is now known as Rue Victor Hugo, which stretches from the Quai de Quebec on the banks of the Dordogne about 500 metres up to the main road. In 1790 the residents of this street were listed as merchants, tailors, locksmiths, shoemakers, hatters, coopers, hairdressers, bakers, nailers, and cutlers.

Today, Rue Victor Hugo is still a busy shopping precinct, at least a few blocks up from the river, away from the warehouses. Jeanne's home is probably still here somewhere, remodelled and repurposed, perhaps with an added third floor, a resurfaced façade, new ironwork and windows. It is impossible for me to know the ages of these buildings, aside from a few conspicuous half-timbered upper floors characteristic of the Middle Ages or some early nineteenth-century towers, both Romanesque and Romantic. Maybe the garden and the well are still there, tucked behind the shopfronts, or perhaps they are covered over with a haphazard patchwork of tiled roofs built every which way in the spaces between the older buildings.

In any case, it is enough to know that this is the street Jeanne walked on, these are the buildings Jeanne passed and this is the town she called home with her husband. Here, she and Dubernat paid their taxes, went to François' wedding, attended a baptism and became godparents to the son of Jean Dumas, a local merchant and tanner. Ordinary citizens living an unremarkable life.

I had not expected to find this level of detail about Jeanne's return to France. Before I arrived in France for this research trip, I had come across a bibliography of all the known material about Jeanne's life in Dordogne compiled by members of the Société Botanique du Périgord. I had tried to contact them before I left, but the phone numbers were no longer connected and the email addresses didn't seem to work. Christèle sent an enquiry to an unpromisingly generic 'jeannebaret' email address, but we did not have much hope of hearing anything back.

The bibliography did at least identify where Jeanne had lived and where she was buried. It also hinted, intriguingly, at the existence of a will.

'Everyone had to write a will then,' Carol had assured me when we were in Paris. 'It will be somewhere in the archives.'

But I cannot find any wills in the online archives for Dordogne, only the parish registers and the cadastral plans.

I expect the wills are among the notarial archives, catalogued by lawyer and year. These are not digitised or online; searching them is a task that needs to be done locally. Not for the first time, I rue my decision to research a topic where the resources are located half a world away in a language of which I have only the most rudimentary grasp. Even when I was in France, I had a limited time for archival work, which, by its nature relies on the almost serendipitous discovery of treasures when and where you are least looking for them.

It was some time after I had returned to Australia that we received an email response from Sophie Miquel. She was working elsewhere at the time Christèle emailed but wonders if we have seen her latest paper on Jeanne Barret, published in the *Cahier des Amis de Sainte Foy*, kindly attached? I had not.

The papers are a goldmine. The archival research is painstaking and detailed. Here are the traces of Jeanne's later life that I had long suspected could be found. Fragments of maps, notarial notices, parish registers and testaments weave together a forgotten life. Unknown for centuries, not because it was hidden, but because no-one had put the effort into looking. I am reminded that, in science, it is very often the local birdwatchers, home astronomers and fossickers who notice the re-emergence of a long vanished species, or photograph a new star or comet, or locate a new fossil. Professional research is not so much about 'standing on the shoulders of giants' as being swept along in a great collective collaborative effort generated by the actions of countless and sometimes nameless individuals. This chapter is largely based on the work of Sophie Miquel and Nicolle Maguet from the Société Botanique du Périgord, and genealogists like Alain Morel. Without their work, we would still know next to nothing about the later life of Jeanne Barret.

The house in Sainte-Foy was not the only property Jeanne and her husband purchased. Just a few months after they bought their first house for 2000 livres, in January of 1776, they bought land in Les Graves near Saint-Aulaye. This was a much larger property, with a house, barn and outbuildings, gardens, farmland, uncultivated land and a hemp field. The land covered 16 hectares and cost the significant sum of 21,000 livres. They paid for it in cash – in gold louis and silver crowns.

That is a lot of money. In the records, Dubernat is sometimes described as a merchant or trader, and a successful one it seems. But it was not through Dubernat's efforts that this wealth had been generated, it was from Jeanne's. Under French law, a woman's property automatically became her husband's unless the husband stipulated otherwise. Although it was Dubernat's name on the purchase of the Les Graves property, he made it clear, in a declaration, that these properties had been purchased jointly by him and his wife. Jeanne was ever careful to protect her assets and her wealth.

'The goods are from common funds to one and the other,' recorded the notary. 'In as much as, Dame Barret provided half of the cash to make the acquisition and the Sieur Dubernat said that he will be guided by his sense of honour in wishing to pay tribute to the truth and agreed that his wife owned an equal share of the property.'

I had almost given up hope of receiving the inventory of goods from Commerson's Paris apartment from the National Archives. But it duly appears in my inbox, one page at a time, ready to download. The notary Regnault has kept all the records in the matter of Commerson's estate, from the moment they open the apartment in August 1774 until Jeanne and her husband arrive in Paris on 3 April 1776. Jeanne had come to Paris to claim her share of Commerson's estate.

Immediately after Commerson's death in Mauritius Jeanne had written to Vachier, who replied to assure her that that he would keep the money owed to her (including from the sale of the furniture) until she returned. And so in April she met with Vachier and the notary to claim her share of her inheritance from Commerson's will. His property had long been sold, but she duly received some 465 livres. On the document her signature is bold and confident, with a large sweeping B. Jeanne was no longer a penniless peasant at the mercy of the rich and well educated. She knew what she was due and how to claim what was rightfully hers.

Vachier's inventory does not just document the furniture, but also Jeanne's clothes. I lay out the items I can decipher from the handwritten text: six blouses, a pair of faux stone earrings, petticoats (one of brown crepe), skirts, a pair of white silk stockings, chiffon lace cuffs, two round bonnets, various red-chequered aprons of various sizes and wear, and a plaid waist apron for special occasions.

This list brings to mind memories of the cut-out paper dolls I had as a child, with various paper costumes and outfits to dress them in. It reminds me too, of the images of seamstresses in Diderot's *Encyclopédia*. Below the engravings of the women in the shop there are often outlines of the clothes they made, laid out flat as if providing patterns for reproduction.

I can imagine Jeanne in a dark skirt and white shirt with lace cuffs, a neat red-chequered apron and round bonnet. I like the idea that I can finally dress her in her own clothes, not those of my imagination. It is a minor detail but it feels oddly important.

I wonder if Jeanne also travelled to her home town of La Comelle, or to see her sister in Toulon-sur-Arroux when she went to Paris. I wonder, too, if Jeanne's sister had been ill and if this played any part in her decision to return to France. Whatever the connection, Jeanne's sister died on 12 March 1777 in Toulon-sur-Arroux, seventeen months after Jeanne's return. I hope they saw each other before she died. Jeanne was obviously close to her sister's family.

Jeanne's sister's son – one of her surviving nephews, Romain Gigon – was 22 years old when his mother died. His half-sister, Françoise Lanoiselée, was about sixteen years old. I don't know when they moved, but they were both living near their aunt in the Dordogne by 1784. They stayed for the rest of their lives – marrying, having children and being included in Jeanne's will as her heirs.

Dubernat's family too, benefited from the traveller's return. His siblings lived on the farm at Les Graves while he and Jeanne stayed in the house in Sainte-Foy-la-Grande. Within a few years a new 'mansion or the great house' was built on the property. It seems likely that after Jeanne and Dubernat sold their house in Sainte-Foy in 1784 (for a tidy profit of 3300 livres), they moved to the great house in Les Graves, where they lived with various nieces from both sides of the family. Their extended family stretched around them in a small radius, just as Jeanne's had in La Comelle. None of them lived more than a few miles from each other, and their signatures adorn the various births, marriages and official records of each other over the decades.

I am relieved to find that the property we visited in Les Graves was, indeed, the cluster of farmhouses where Jeanne lived. The local researchers found the cadastral maps of Les Graves, and have pinpointed the exact boundaries of each parcel of land and its use. I can't be sure if the buildings were exactly the same ones, or if they were new, or simply like the metaphorical age-old axe whose shaft and blade have each been replaced on multiple occasions over a regenerating life. Architectural fashions don't change much in the country. It would take a more experienced and knowledgeable eye than mine to distinguish the traces of the different centuries – but they are clearly still there, written into the fabric and foundations of the buildings as much as into the furrows and fields of the land. You can still see the outline of the Barret-Dubernat lands on an aerial photograph of Les Graves, the turn of a creek line, the route of a new road, the long thin ploughing of a field following ancient lines, unmarked by fence or visible boundary but held in place by the habits of centuries.

Inside Jeanne's house, it seems, there were no mementos of her voyage. Among an inventory of her chattels, her movable goods, there was no collection of shells, no polished coco-de-mer on the mantelpiece, no books about plants or voyages. No books at all, in fact, not a single one. Which makes me wonder whether she could read and write after all.

Several large timber cupboards lined her bedroom. One was lockable, suggesting goods of value, like the two yellow copper basins, each valued at over 30 francs. But the room was dominated by the large cherrywood armoire stacked with linens, new and used: 52 women's shirts, 24 bedsheets, 52 towels, 30 tablecloths, twelve aprons, nearly 100 handkerchiefs, 40 bonnets, eleven pairs of pockets, six pairs of woollen stockings and two bodices. A striking difference from the six shirts and two bonnets she had left in Paris all those years ago.

A cherrywood bed, hung with curtains and a burnt-wood 'sky', was laden with comfortable mattresses, heavy quilts stuffed with goose feathers and patchwork-quilted bedspreads. One of these was made from several different Indian fabrics bordered with fine 'partridge eye' embroidery. I wonder if the Indian fabrics were a relic of her time in Mauritius, on the trade route from India. But India was a powerhouse of textile production and the biggest exporter to the world before England looted its wealth and independence, so even this was nothing unexpected for the times.

Elsewhere in the house was an abundance of food and wine. It seems they probably ran a wine business, judging from the number of vats in the cellar. In the attic were sacks of wheat, rye, beans and gesse or chickling vetch. There was a profusion of bacon and ham too, from the pigs in the stables. All the staples of good traditional peasant food, with nothing exotic in sight. Jeanne was living a affluent, well-off life – never short of food, clothes or a comfortable bed, all things in short supply on a ship, or in a peasant's home. I don't think she missed her life at sea or her childhood.

'The voyages of the seas,' Commerson once wrote, 'always hard and weary, which so quickly wear away the fabric of life, carry with them an indefinable charm. This mixture of privation and abundance, of idleness and activity, of calm and tempest, this immensity of the seas and these so pompous regions of the torrid zone, leave only deep impressions; after them, domestic happiness is monotonous; and this is what causes so many travellers to break the promise they have given a hundred times to no longer venture on the waves.'

I think about living on a boat, sometimes: the particular lightness of air, sounds floating over water, the sense of drift, the minimalism and smallness, self-contained, unburdened, untethered and free. A tiny kingdom of one's own. Unlike Jeanne, my bookshelves, paintings and shells continue to betray my maritime fascinations. But I remind myself of a promise I made when I returned from New Guinea, that if I ever considered returning to sea, I would first go, one after the other, on the most hair-raising showground rides. My stomach turns of its own accord just thinking about it. It has its own visceral memories.

Commerson may well have been right about the irrational attraction of adventures at sea. But his place in history would have been more assured if he had travelled less and finished his books, completed his work. Perhaps a more mundane domestic happiness would have secured his legacy, served our knowledge and the protection of Mascarene biology better. But even if I have no real interest in returning to a life at sea, even if Jeanne seems to have retained not a single memento, it does not mean such voyages were not significant. They made me a biologist and a writer. They shaped my beliefs and my habits even though, like Jeanne, I have no great need to return there and have happily returned to more terrestrial occupations.

Jeanne had not forgotten her journey, and she did not expect others to have forgotten her contribution to it either. In 1785, she was awarded a naval pension for her services.

> Jeanne Barré, by means of a disguise, circumnavigated the globe on one of the vessels commanded by M. de Bougainville. She devoted herself in particular to assisting M. de Commerson, doctor and botanist, and shared with great courage the labours and dangers of this savant. Her behaviour was exemplary and M. de Bougainville refers to it with all due credit. When M. de Commerson died, this person, whose sex had been recognised, married one Dubernat, formerly a non-commissioned officer in the Royal-Comtois Regiment.
>
> Today, she (née Barré) and her husband, having reached an age that brings infirmities with it and no longer able to earn their living, His Lordship has been gracious enough to grant to this extraordinary woman a pension of two hundred livres a year to be drawn from the fund for

> invalid servicemen, and this pension shall be payable from 1 January 1785.

When the payments from Paris were not forthcoming, Jeanne assigned power of attorney to Parisian colleagues to secure the back payments due to her.

The image of Jeanne Barret in her later life refines and remodels my image of her on the voyage. For much of this story, she has been elusive and difficult to characterise. The sheer brevity of the records, refracted and distorted through a male gaze and the absence of her own voice, have made it challenging to build a picture of her. But the actions of the mature Jeanne consolidate all the hints and possibilities of her youth.

She comes into focus, solid and unwavering, a woman who knew her rights, and claimed and defended them. A strong soul in a far from frail machine. She was a quiet, unassuming woman who did not blow her own trumpet but who also stood her ground. She was strong, agile, courageous, determined, hardworking, tireless and determined – indefatigable, in fact. She was a serious woman with an intense curiosity about the world. A compassionate, caring woman – immensely loyal, honest and generous. She behaved decorously, with scrupulous correctness, and yet unabashedly broke convention to attain her goals. She was clever, a quick learner, a good and thrifty manager. I imagine her as Colette once described her own mother Sido: curious, eclectic, independent, self-contained, staunch, devoted to those she cared for and the provincial community she lived in.

We need to stop thinking of Jeanne as a figure eternally standing in Commerson's shadow. In her later life, as in her early years, she was not alone, but belonged to a close and caring family. She was an independent woman who determined the course of her own life, before and after, and was surrounded by a family she took care of, and who cared for her – nieces, nephews, great-nieces and great-nephews.

For many years, biographers claimed that on her death, Jeanne Barret bequeathed all her worldly goods to Commerson's son, Archambaud, and his family. It's a touching story, and one that fits well with the model devoted servant. The story originates from Commerson's great-nephew, J. B. Jauffred, of Châtillon-les-Dombes, who said that she 'finished her days at Châtillon and, by way of remembrance and veneration for her former master, she left all she possessed to the natural heirs of the famous botanist'.

But Jeanne did not return to live in Commerson's birth town, did not outlive her husband and did not leave all she possessed to Archambaud. Perhaps there is a grain of truth to the story. Perhaps she sent some items belonging to Commerson to Archambaud after she returned to France and this kind act has magnified in family history over time. But in the three wills signed by Jeanne Barret in the Dordogne archives, there is no mention of Commerson or his heirs. There is no mention of their son, Aimé Bonnefoy, either. She left her estate to husband Jean Dubernat and to her various nieces and nephews, the children and grandchildren of her brother Pierre and sister Jeanne, as well as a bequest to the poor.

Her signature had become increasingly shaky and uncertain on these wills, as if she had forgotten how to shape the letters or how to spell her own name. Perhaps the trials of her youth were catching up to her. But she lived longer than either of her parents, her siblings, most of her nieces and nephews and probably both of her own children, and seen more than any of them could ever have imagined.

Jeanne died in her home on 5 August 1807 on the stroke of midnight at the age of 67.

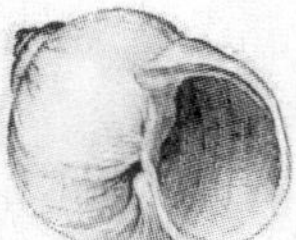

EPILOGUE

As we wind our way into Saint-Aulaye itself towards the Dordogne River, the scenery shifts from the flat monotony of farmland. The riverbank ahead is shaded with broad leafy trees along its grassy bank. The church spire and the wildflower-filled graveyard welcome visitors with an open gate. Along the waterfront the houses open their doors and windows, and pretty gardens sprawl over fences and archways, as if the river itself draws the inhabitants out of doors and some intangible current entices them out onto the street, down the bank and along the river towards the open sea and the world beyond.

It is not easy to find Jeanne's gravestone in the summer grass left long to encourage wildlife and biodiversity. Tiny white flowers scatter through the seeding grass like stars, while bursts of bright red-orange poppies punctuate the swathe of yellow-green. Botanist, explorer, adventurer – the first woman to circumnavigate the world disguised as a man. I'm not really sure what I expect to find. A tropical shell left on the gravestone? A bougainvillea flowering in all its vibrant brilliance up the church wall? Some grand epitaph telling her story?

The church at Saint-Aulaye perches right on the riverbank of the Dordogne River. A statue of Our Lady of Lourdes watches over those who pass by on the river. We walk a circuit, wading through the long summer grass, suppressing Australian anxieties about snakes. We triangulate off the surrounding buildings, looking for landmarks.

And finally, we find it, a slab with a rectangular carved headstone and an ornate metal cross above. It's no longer possible to read the inscription, but there is a small handwritten label, from a local committee, designating the site as Jeanne Barret's grave. Nature provides its own adornments, surrounding the grave with delicate riffs of pink, white, orange, yellow and purple flowers waving across a lush green sea of grass. I'm not sure if this is her original gravestone, or even the original gravesite. I don't know if someone has replaced it over the 212 years since her death, or just decided that this one will do to designate her burial place. I can't even be sure that her bones lie here and whether it even matters.

We wander down to the river's edge, where the shade of large trees provides relief from the hot sun. The water runs clear and fast over the shallows, rippling and turning just beneath the surface. Islands of floating vegetation scatter the full width of the river. I suspect its proliferation is a sign of excess nutrients, an unbalanced system impacted by too much agricultural run-off, but the scattering of its white flowers reminds me of stars reflected in water at night and of Tolkien's simbelmynë, which blossoms in all seasons of the year and grows where the dead take their rest.

It is a fitting resting place for one who belongs in the annals of the great women of the world. She did not aspire to greatness or glory: 'she had no dreams of being praised above other women'. She merely set out to live her life on her own terms – whether in discovery on the world's oceans, climbing mountains in the Strait of Magellan, or navigating social mores in Tahiti or in the fields of the French countryside. For now, she rests in a small cemetery in rural France, both an ordinary and an extraordinary woman, one who never asked to be remembered but perhaps should have been remembered better, one whose life is an inspiration and a model for those who long for adventure and for those who crave a quiet and productive life. This story has revealed where this remarkable woman came from, how she lived and died. That she is more than just a footnote to other men's lives. She has her own life story to tell.

ACKNOWLEDGEMENTS

THIS BOOK COULD NOT have been completed without the assistance of a great many people. First and foremost I must thank my colleagues, Dr Christèle Maizonniaux and Professor Carol Harrison, for their ongoing collaboration, for generously sharing their scholarly expertise and language skills, and most importantly, for their company on my research trips in France. Thank you also to Miles Dodd and Jocelyn Probert for generously hosting and coordinating my research in Burgundy where Jeanne Barret was born. Research is much more fun in such good company.

Given that many of the resources for this book are in languages other than English, I am greatly beholden to Patricia Komarower for some of the French translations, Ian Gibbins for German, and my daughter Rachel Nicholls for Latin. Equally important, if only to keep me sane while writing, was the support and enthusiasm of my wonderful writing groups and book clubs, and the advice and feedback of readers Sara King and John Clode.

I must also acknowledge the assistance of all the museum, library, herbarium and archive staff without whose work and assistance this book would not have been possible. As always I am grateful for the research support of Flinders University and my colleagues in the College of Humanities, Arts and Social Sciences as well as research funding from the South Australian government through an Individual Makers and Presenters grant from Arts SA and a Brittany Scholarship from the Department of Premier and Cabinet.

Like all research, my efforts are based on a foundation laid by others, as acknowledged in the references. In addition, I would like to thank John Dunmore for his large body of work on French voyages in the Pacific, particularly the Bougainville voyage. I am also grateful to Jean-Marie Huron for assisting me with archival research in Mauritius, as well as Sophie Miquel for directing me to her new published work on Jeanne Barret in the Dordogne.

Thanks to my commissioning editor Mathilda Imlah for her patient encouragement and to my agent Jenny Darling for her ever-constant support. I feel blessed to be able to turn this manuscript over to the talented and professional publishing crew at Picador, particularly Georgia Douglas and Belinda Huang, as well as Nicola Young, Elena Gomez, the team at Post and cover designer Debra Billson, and see them transform it into a beautiful book.

And finally, to my family who have listened to far more about this book, far more often, than anyone ever should – thanks for being there for me.

NOTES

p1 'A frail machine . . .' refers to Diderot's 1796 description of Jeanne Barret.

1. The Call of the Sea

p6 Archaeological and literary evidence of Gudrid Thorbjarnardóttir's journey, Brown 2007; history of woman at sea, Creighton and Norling 1996; Druett 2000; Snell *et al.* 2008 and Wheelwright 1989. **p8** Mauritius was known as Île de France during this time. I have generally used modern rather than contemporaneous names throughout to avoid confusion. The early references to Jeanne include Bougainville, 1771; Lalande 1775 and Lamarck 1804 p44. **p9** 'neither shame nor . . .' and 'the act of . . .' Commerson 1769. 'These frail machines . . .' Diderot 1796. Commerson's will was written in 1766 (as reproduced in Monnier 1993 pp42–46) but was made public by an unknown publisher, Commerson 1774 p13. This version is available at books.google.com.au/books?id=w6qMVt9gJeoC&dq=testament singulier de M. Commerson&pg=PA1 – v=onepage&q=testament singulier de M. Commerson&f=false. 'genius for a little romance' and 'was very attached . . .' Bachaumont 1784 p159. **p10** An example of early romanticised or fictionalised accounts of Jeanne's journey include La-Croix 1788; Vinson 1874 and Oliver 1909. More recent biographical articles about either Jeanne specifically, or within articles on Commerson, include Role 1973; Édouard 1974; Christinat 1995; Schiebinger 2003; Lignereux 2004 and Jolinon 2005. 'preferred to leave . . .' Dussourd 1987 p10. **p11** 'Most scholars are . . .' Finney 1979 p12.

2. An Inauspicious Beginning

p15 Regional French culture and history, Robb 2007; historical weather Marusek 2010. **pp16–17** Historical and archaeological details about La Comelle, Niaux 2000. Details of famine in eighteenth-century France, Kaplan 1982; deaths in Poitou, Marusek 2010 p219; Morvan birth and baptism rituals, Bareille 2019. 'On the twenty-seventh . . .' Acte Naissance, Jeanne Baret, 27 Jul 1740, Archives Municipales Saône-et-Loire, Comelle (La), Baptêmes, Mariages, Sépultures 1736–1749, 4E dépôt 142/2, p55, www.archives71.fr/ark:/60535/s005139982dac457/5139cad60fdde; see Gilles Pacaud for a recent discussion (March 2020) of early records associated with Jeanne Barret and her family www.jeanne-barret-tourdumonde.fr. For the purposes of this work, I have primarily relied on archival documents that I found or could verify from their sources or manuscripts. **p18** Details of Frank Jaunay's early career in Australia can be found in *The Argus* (Melbourne), 2 Mar 1894, p5; 'Our wine and brandy industries: the Chateau Tanunda', *Supplement to the Chronicle* (Adelaide), 12 Sep 1896 pp97–103; 'Adelaide Licensing Bench', *Adelaide Observer*, 17 Sep 1898 p14; and *The Register*, 10 Sep 1902 p3. **p19** For a discussion of these French voyages of exploration to Australia see Clode 2007/2018 and Bloomfield 2012. **p20** Historical demographics, Henry and Blayo 1975; cultural traditions in France, Robb 2007; rates of infant mortality in 1740, Kaplan 1982 p64; descriptions of poverty in the Morvan, Pierre Goubert and Sébastien Le Prestre Vauban, quoted in Dunmore 2002a pp12, 14–15, also Root 1987 p11; details on Morvan peasant diets, Durand and Wiethold 2014 p417; 'mowing, harvesting, threshing . . .' Vauban quoted in Dunmore 2002a p13. **p21** Details on the discussion of Pochard as a Huguenot name, Ridley 2011 p36. Data on the distribution of French surnames across regions and times from www.geopatronyme.com/cgi-bin/carte/nomcarte.cgi?nom=pauchard&submit=Valider&client=cdip. 'One the fourth . . .' Acte de Mort, Jeanne Pauchard, 4 Nov 1741, Archives Municipales Saône-et-Loire, Comelle (La), Baptêmes, Mariages, Sépultures 1736–1749, 4E 142/2, p71, www.archives71.fr/ark:/60535/s005139982dac457/5139cad61f7e0. **p22** The postcard of the church at La Comelle, Niaux 2000. **p24** 'spend their days . . .' Guy Thuillier quoted in Robb, 2007 p76. **p26** Acte de Mort, Antoinette Mangematise et Simon Baret, 4 Nov 1745, Archives Municipales Saône-et-Loire, Comelle (La), Baptêmes, Mariages, Sépultures 1736–1749, 4E 142/2, p123, www.archives71.fr/ark:/60535/s005139982dac457/5139cadb92b8b. **p27** Acte de Mort, Jeanne Teuvenot, 17 Nov 1747, Archives Municipales Saône-et-Loire, Comelle (La), Baptêmes, Mariages, Sépultures

1736–1749, 4E 142/2, p150, www.archives71.fr/ark:/60535/s005139982dac457/5139cad-bad16e. 'On the 16th . . .' Acte de Mort, Jean Baret, 16 Dec 1755, Archives Municipales Saône-et-Loire, Comelle (La), Baptêmes, Mariages, Sépultures 1750–1761, 4E 142/2, p70, www.archives71.fr/ark:/60535/s005139982dae397/5139cadc1f7ca. 'The 12th Feb . . .' Acte Naissance, Pierre Barret, 12 Feb 1734, Archives Municipales Saône-et-Loire, Comelle (La), Baptêmes, Mariages, Sépultures 1718–1735 E dépôt 2587, p56, www.archives71.fr/ark:/60535/s0051d6b717badee/51d6b78dec95c. **p28** 'being chained to . . .' Colette 1929/1968 p38.

3. Making her Mark

p31 Details and age of Jeanne's older sister, Acte de Décès, Jeanne Barret (elder), 12 Mar 1777, Archives Municipales Saône-et-Loire, Rosiére Toulon-sur-Arroux, Baptêmes, Mariages, Sépultures 1748–1780 E dépôt 6173–6175, p178, www.archives71.fr/ark:/60535/s00513996d55986e/5139b293bc95f. **p32** See details of the French genealogist's work at Alain Morel, gw.geneanet.org/alainjm33?lang=en&pz=alain+jean+michel&nz=morel&p=-francoise&n=lanoiselee. Acte de Marriage, Jeanne Barret (elder) and Antoine Gigon, 12 Feb 1754, Archives Municipales Saône-et-Loire, Dettey, Baptêmes, Mariages, Sépultures 1753–1764 E Dépôt 1446, p9, www.archives71.fr/ark:/60535/s005139982bc79eb/5139c9cc49afe. Information on Morvan wedding traditions, Bareille 2019. Acte Naissance, Romain Gigon, 1 Jan 1755, Archives Municipales Saône-et-Loire, Thil-sur-Arroux, Baptêmes, Mariages, Sépultures 1748–1778 Collection communale, p66, www.archives71.fr/ark:/60535/s0051399833ca8a4/5139ced599c15. **p33** In later years, Jeanne's sister (and nephew) would be buried at Rosiéres, leaving her surviving niece Françoise an orphan at the age of seventeen. **p34** Description of Commerson, Lalande 1775; Cap 1861 and Morel 2012a p2. Note that the measurements have been converted from French feet (which were larger than English feet). **p35** 'She is something . . .' Dunmore 2002a p26; theory of Jeanne as a herbwoman, Ridley 2011 p34. **p36** 'Table des plantes médicamenteuses' Muséum National d'Histoire Naturelle Bibliotheque, Commerson Archives, Ms 884; typical medical practices including herbal remedies of the time, Bonnot 1787, who was one of Commerson's successors at Toulon-sur-Arroux. **p37** Analysis of Jeanne's name, Ghabut 1998 pp61–62. **p38** 'Short of stature . . .' Vivez, and 'neither ugly nor plain', Bougainville, Dunmore 2002b pp223 and 97. **p39** 'Jeanne Barret, adult . . .' Declaration de Grossesse, Jeanne Barret, 22 Aug 1764, Archives Municipales Saône-et-Loire, Digoin, Actes, notaire Labelonye, 3E 22802. Jeanne describes herself as a *domestique*, a servant, although Commerson described her as his *gouvernante*. A *gouvernante* can be a nursemaid, a housekeeper or a governess. This may be taken to mean that Jeanne cared for Commerson's son (Crestey 2011). This word also means, however, the female housekeeper of a bachelor or widow. The witnesses to Jeanne's declaration of pregnancy were Hugues Maynaud de Bizefranc esquire, lord of Lavaux, Cypierre and Saint-Louis, known as a generous humanitarian and father of famed anti-slavery campaigner, and Louis-Henry-Alexandre Laligant, a 35-year-old medical practitioner and graduate from the same Montpellier medical school as Philibert Commerson (Mémoires de la Sociéte Éduenne, New Series, V10 (Imprimerie de Michel de Jussieu: Autun, 1859) p96, archive.org/stream/mmoiresdelasoci26frangoog/mmoiresdelasoci26frangoog_djvu.txt. See also Geneanet profile gw.geneanet.org/cal59?lang=en&m=N&v=laligant. Dussourd was an expert in the rural history of central France, see Dussourd 1968 and 1987. **p40** 'less valued either . . .' Commerson, Dussourd 1987 p23. 'He often spent . . .' Beau to Minister Turgot, 1774, Archives Nationales, AN Col E 89, Philibert Commerson files. Accessed at www.pierre-poivre.fr/doc-74-an-c.pdf. The path of the solar eclipse across central France compiled by Nicole-Reine Lepaute 1762 (engraved by Mme Lattre and Elisabeth Clare Tardieu Lepaute), 'Passage de L'ombre de la Lune au travers de l'Europe dans l'Eclipse de Soleil Centrale et annulaire qui s'observent le 1er. Avril 1764' (Lattre: Paris). Commerson mentions Lepaute as a friend in a letter to Beau, 20 Oct 1766, Cap 1861 p84. **p41** 'I have lost . . .' Commerson to Bernard, 8 Jun 1762, Cap 1861 p80. Details of the plant Commerson named after his wife (now *Polycardia phyllanthoides*), Cap 1861 p12. **pp41–42** Obligatory declaration of pregnancy, decreed by Henri II, Feb 1556, Pablo Briand, 'Les Enfants Trouvés', www.archivosgenbriand.com/chron_obliv_fr.html. Commerson's will (1764) reveals that Jeanne had been working for him since at least 6 September 1764, just fifteen days after she declared her pregnancy in Digoin. **p43** Pre-scientific beliefs about fossils, Duffin 2008; Taylor 1998.

4. A Woman of the World

pp45–46 Descriptions of journey from Burgundy to Paris, Anon. 1792. **p47** The church originally dedicated to Sainte-Geneviève became the Panthéon. Description of the Paris apartment, Yolande Zephirin, Actes du Colloque, Commerson 1973. *Cahiers du Centre Universitaire de Réunion*, Dussourd 1987 p22 and from the drawing by Louis-Nicolas Lespinasse, 1788, *Vue intérieure de Paris, prise du belvédère de M. Fornelle, rue des Boulangers St Victor,* Paris Musées D.5366. Details of Morvan wet nurses, Drake 1940. **p52** Labillardière's Paris residence, Duyker 2003 p23; rednecked wallabies *Macropus rufogriseus* were taken back to France on the Baudin voyage from King Island; 'the father of . . .' Mayr 1982 p330. **p53** 'The sorrow he . . .' Cap 1861 p12; 'The truth you . . .' and 'Although no-one has . . .' Commerson to Beau, in Dussourd 1987 pp23–24. **p54** The impact of *lettres de cachet,* Farge 1993 p78. 'I kiss you . . .' Commerson to Beau in Cap 1861 p85. **p55** 'put to the . . .' with details from the register of entries to the hospice, 14 Jan 1765, Archives of the City of Paris, N° 152, in Dussourd 1987 p24. **p57** Parish maps, J. Junié and L. Wuhrer 'Plan des paroisses de Paris' 1768, Norman B. Leventhal Map & Education Center, collections.leventhalmap.org/search/commonwealth:6w924q18p. **p58** 'To see something . . .' Jonathan E. Kolby, 'Everything is on fire' 15 Apr 2019, twitter.com/MyFrogCroaked/status/1117924267864227841. **p59** Details of mollusc extinctions, Régnier *et al.* 2009; 'offspring of a . . .' Hugo 1831 Ch1; history of foundling children, Seth 2012; 'After pulling the . . .' Delrieu 1831. **pp59–60** 'I have been . . .' Farge 1993 p39 fn18. Rousseau's abandoned children, Mendham 2015. Gilles Pacaud suggests that Jean-Pierre Barret may not have been Jeanne's son, '5–7 Jean Pierre Baret n'est pas l'enfant de Jeanne Barret et P. Commerson' www.jeanne-barret-tourdumonde.fr **p61** Farge 1993 pp52–53; Jean-Pierre's death, Dussourd 1987 p24; also the records of admissions for assisted children, 'Répertoire du Registre matricule des Enfants Trouvés', 1765 p9, Paris Archives D2HET 8.; women in the French workforce, Sheridan 2009; peasant women's canal building expertise, Mukerji 2008. **p62** Women in early science, Gelbart 2016. **p63** 'It is one . . .' Commerson to Beau, 20 Oct 1966, Cap 1861 p84.; 'all the items . . .' César Gabriel de Choiseul-Praslin, Secretary of State for the Navy, to Commerson, Cap 1861 p91. **pp64–65** 'They tell me . . .' and 'I have been . . .' Commerson to Jean Bernard, Jan 1767, Cap 1861 p81; 'I bequeath to . . .' Commerson 1774, translation from Dunmore 2002a pp43–44.

5. A New Commission

p70 'tormented with an . . .' Melville 1892 p12; 'piqued her interest . . .' Bougainville, Dunmore 2002b p97. **p71** Lapérouse's library, Smeaton 1988; 'induced to go . . .' Flinders 1814b p178, see also Downing 2014. **p72** Polynesian understanding of the Pacific is evidenced by Tupaia's map, which has only recently been better understood, see Eckstein and Schwarz 2018. **p73** Discussion of an indigenous Polynesian history, Hau'ofa 2008 pp60–79. **pp73–74** Biographical details of Bougainville's life, Le Brun 2019; Dunmore 2015. **p75** 'The maritime movement . . .' Verne 1890. **p77** The innovation to the fifth-rate frigates described at 'Frégate de 12', Histoire de Frégates, www.histoire-de-fregates.com/contexte/8-histoire-de-la-fregate/40-fregate-de-douze. For the *Boudeuse*'s crew capacity, Demerliac 1995, see threedecks.org/index.php?display_type=show_ship&id=11140. **p78** Ornament designs for the *Boudeuse*'s stern and bow, 'Dessin d'ornement de bouteille de proue de Lubet', l'Etat Général des Fonds des Archives Nationales, sous-série Marine D1 – constructions navales – (référence: D1 68 fol 7 cl 7295). See lesamisdebougainville.wifeo.com/fregates.php. **pp79–80** Nicolas-Pierre Duclos-Guyot was from Saint-Malo, Dunmore 2002b pxxvi. Biographical details of other officers and passengers, Dunmore 2002b. **p82** 'Glory requires, like . . .' Commerson, Jenkins 1978 p47.

6. Fitting Out for the Voyage

p85 Details of French construction methods, Peters 2013. **p86** 'us in funny . . .' Dening 1997 p422. **p87** 'the clumsiness of . . .' Commerson to Beau, 23 Dec 1766, Cap 1861 p94; 'A woman shall . . .' Deuteronomy 22:5; female cross-dressing in the *ancien régime*, Pellegrin 1999. **p88** Cases of cross-dressing, 'with no female . . .' and 'she was a . . .' Farge 1993 pp152, 158. Hannah Snell, *The Female Soldier or The Surprising Life and Adventures of Hannah Snell* (R.

Walker: London, 1750); Mary Lacy, *The History of the Female Shipwright* (M. Lewis: London, 1773); and Mary Anne Talbot, *Life and Surprising Adventures of Mary Ann Talbot in the Name of John Taylor* (R. S. Kirby: London, 1809), republished in Snell *et al.* 2008. The case of 'William Brown', MacMasters 2015. **p89** 'an infinity of . . .' Commerson to Beau, 23 Dec 1766, Cap 1861 p95. **p90** 'The Rochefort Intendant . . .' Bougainville, Dunmore 2002a p46. **pp90–92** Details of the *Étoile*, 'Plan de l'Etoile au SHD (Service historique de la défense) de Toulon' 1L 442, see lesamisdebougainville.wifeo.com/fregates.php; descriptions of Rochefort, 'Plan de la ville et port de Rochefort', 7 Oct 1761, Bibliothèque Nationale de France, Département Cartes et plans, GE C-908, ark:/12148/btv1b53025047b and the 1.65 by 2.63 metre painting by Claude-Joseph Vernet 1762, *Vue du port de Rochefort, prise du magasin des Colonies*, Musée National de la Marine. **p94** 'I have given . . .' Bernardin de Saint-Pierre 1775 p8. **pp95–96** Jeanne's image was probably published in Anon. 1806 p204, although I have been unable to locate a copy of this book with the plate included. The artist was probably Giuseppe dall'Acqua (1760–1829) rather than his more famous father, Cristoforo; see Lodi 1985. **p97** 'Before leaving the . . .' and 'I was extremely . . .' Rose de Freycinet 1817, Bassett 1962 pp6–7; 'the most beautiful . . .' Commerson to Beau, 23 Dec 1766, Cap 1861 p94. **pp98–99** Biographical details of Vivez, Dunmore 2002b pxxxviii and Dunmore 2015 p287; biographical details of Pierre Duclos-Guyot, Dunmore 2002b pxxx; 'A captain in . . .' Masefield 1937 p56. **pp99–100** Role of servants, McBride 1974; roles of ships boys and servants, Pietsch 2004; Jones 2016. **p100** 'I must warn . . .' Commerson to Beau, 23 Dec 1766, Cap 1861 p94.

7. All at Sea

pp103–104 All the early details of the voyage, Commerson's journal, 'Mémoires pour servir à l'histoire du voyage fait autour du monde fait par les vaisseaux du Roi La *Boudeuse* et l'*Étoile*', Taillemite 1977 V2 pp421–25; digitised version available from the Manuscrits de la Bibliothéque du Muséum National d'Histoire Naturelle, Ms2214, 1er cahier. **p104** 'I am no . . .' Commerson, Dussourd 1987 p27. **p105** 'Eat, eat . . .' Commerson, 2–3 Feb 1767, Taillemite 1977 V2 p421. According to the journals of Commerson and later reports from Vivez, both Jeanne and Commerson were ill at the beginning of the voyage, Commerson being the last to recover, suggesting that their movements would have been constrained for the first two weeks, Vivez, Dunmore 2002b p228 and Commerson, Taillemite 1977 V2 pp421–25. **p106** 'liquidly glide on . . .' Melville 1888 p90. **p107** The first account published was Bougainville 1771, and translated into English in 1772. It has been regularly reprinted, in original, abridged, collected and translated forms ever since, and remains in print today. Bougainville's personal journal (along with most of the other journals from the voyage) in French, Taillemite 1977; and in English, Dunmore 2002b. **p108** 'was not proper' Vivez, Dunmore 2002b p228. **p109** Curtained-off servant sleeping quarters, Jones 2016 p121; sharing sleeping quarters with other officers for the Atlantic crossing, Commerson to Beau, 23 Dec 1766, Cap 1861 pp94–95; common practice of sharing beds (and implications for a cross-dressed woman) is vividly described by Mary Lacy, Snell *et al.* 2008; 'rather bantering and . . .' Dunmore 2002b pxxxviii. **p110–112** Details of crossing the line, Commerson, Dunmore 2002b p63. **p116** Bruce Davey interview, ABC891 News, 20 Mar 2019. **p117** 'To push it . . .' Commerson, Taillemite 1977 V2 p424. **p118** 'It had been . . .' and 'But he knew . . .' Saint-Germain, Crestey 2011 p7. **p119** 'Scandalous gossip claims . . .' Vivez, Dunmore 2002a p228; evidence for Vivez treating Commerson, Dunmore 2002b p302 and Commerson to Beau, 7 Dec 1767, Cap 1861 p102; women's health and sexually transmitted diseases, Doig and Sturzer 2014; Hannah Snell writes of removing a bullet, Snell *et al.* 2008. **p120** Brief account of Marie Louise Victoire-Girardin, Duyker and Duyker 2006 pxxv.

8. Gathering Nature's Riches

p122 Laws of the Indies, Siembieda 1996. **p123** 'The first reflection . . .' Bougainville, Forster 1772 p31; Commerson's description of Montevideo, Lalande 1775, as well as the English translation in *The Monthly*; 'the most beautiful . . .' Commerson to Beau, 7 Sep 1767, Cap 1861 p101; 'the same perplexity . . .' Commerson to Beau, 28 May 1767, Cap 1861 p99; description of the naturalist is Charles-Antoine-Gaspard Riche from the journal of Jacques-Malo de la Motte du Portail, Clode 2007 p55 fn270; 'During our stay . . .' Vivez, Taillemite 1977 p238. **p124** These birds were described as being collected by Commerson in Buffon and Cuvier 1831–32 V23 as *Motacilla perspicillata* (now *Hymenops perspicillatus*) p178 and *Alauda rufa* (now *Lessonia rufa*) p54; Darwin's use of servants for collecting work, and 'employed a labourer . . .' Darwin 1887 V1 p63, 50; locality data for the Galapagos finches and Covington's collection, Sulloway 1982. **p125** 'Notwithstanding our smiles . . .' Fitz Roy 1839 pp106–107, Covington 1839 Ch3 fn37 online edition, web.archive.org/web/20030511133343/ austehc.unimelb.edu.au/bsparcs/covingto/chap_3.htm#cite37. **p126** Details on South American shells from Bougainville's voyage, Martens 1872 p53. **p127** The white catfish is *Genidens barbus,* which Lacépède 1801 also described as *Arius commersonii,* which is only found in the Rio de la Plata; estimates of plant specimens taken from the geographical breakdown of specimens listed on JSTOR Global Plants (plants.jstor.org) with Commerson as collector. **p129** Although Jeanne is not directly credited as the collector on any of the specimens she collected, a new plant species has recently been named after her, Tepe *et al.* 2012. **p130** 'I was obliged . . .' Commerson to Beau, 16 Jan 1770, Montessus 1889 p201. **p131** Difficulties with naturalists are illustrated from the d'Entrecasteaux voyage, when the botanist was enraged that the plant presses were removed from the warm, dry great cabin the officers shared and placed in an exposed damp area. D'Entrecasteaux then ordered that they be returned, Duyker and Duyker 2006 p104. **p132** 'The immense forests . . .' Rose de Freycinet at Rio de Janeiro, Bassett 1962 p29; 'M. de Commerson . . .' Bougainville, 5 July 1767, Taillemite 1977 V2 p229; see also Lack 2012 p118 for details of Commerson and Jeanne's work here. **pp132–133** 'You know my . . .' and 'M. de Bougainville . . .' Commerson to Beau, 7 Sep 1767, Cap 1861 p102. **p134** 'In the middle . . .' and 'I am bringing . . .' Commerson to Beau, 7 Sep 1767, Cap 1861 pp101–102, 104. **p136** Descriptions of the reddish-brown birds (now *Funarius rufus commersoni,* or the red ovenbird), Buffon 1799 p476. **p137** Wildlife in the Strait of Magellan, Bougainville, Forster 1772 pp126–27. **p139** 'found the sea . . .' and 'I have seen . . .' Bernardin de Saint-Pierre 1775 pp24–25; details of these floating pelagic ecosystems, Helm 2019. **p140** Descriptions of Commerson's dolphin, *Cephalorhynchus commersonii,* Commerson, Lacépède 1804 V1 p104; descriptions of botanising, Bougainville and Vivez, Dunmore 2002b pp12, 19, 229; and Commerson, Cap 1861 p37; 'Frightful weather . . .' Bougainville, 4 Jan 1767, Dunmore 2002b p27. **p141** In the Strait . . .' Vivez, Dunmore 2002b p229; Caro on leaving the Magellan Strait, 27 Jan 1768, Taillemite 1977 V2 p322; 'We will sail . . .' Commerson to Beau, 7 Sep 1767, Cap 1861 p106.

9. Pacific Voyagers

p144 My first published book was *Killers in Eden*, Clode 2002/2011; La Giraudais changing course mid-Pacific, Vivez, Dunmore 2002b p224. **p145** 'Who the devil . . .' Caro, 21–22 Mar 1768, Dunmore 2002b p200; 'All Europeans are . . .' Lapérouse, Milet-Mureau 1799 V2 p179. **p146** Tuamotuans as master shipwrights, Klem 2017 p4. **p147** 'Lost in blue . . .' Stevenson 1908; *South Sea Tales,* London 1911 pp261–306. **p148** 'infinte numbers', Bougainville, 22 Mar 1768, Taillemite, 1977 V1 p305; research on Tuamotuan birds, Blanvillain *et al.* 2002 (species include Polynesian ground dove *Gallicolumba erythroptera*, Tuamotu sandpiper *Prosobonia cancellata*, atoll fruit-dove *Ptilinopus coralensis*, Tuamotu reed-warbler *Acrocephalus atyphus*, spotless crake *Porzana tabuensis* and bristle-thighed curlew *Numenius tahitiensis)*; impact of cyclones on coral atolls, Duvat *et al.* 2017. **p149** Details of Solomon Island sea level rises, Albert *et al.* 2016; Morton and Stephens 2015. **p150** French nuclear testing in the Pacific, Chrisafis 2013; 'Don't feel afraid . . .' dates from 1920s, probably written by R. McCann and reproduced by Buzza Motto on various sentimental mirrors and prints featuring children and dogs.

10. Revelation

p154 Geological origins of Tahiti, Neall and Trewick 2008; land tortoise dispersal, Cheke *et al.* 2017. **p155** Tahitian plant endemicity, Meyer and Florence 1996; 'Scarcely did we . . .' Bougainville 1771 pp187–88. **p159** 'head or Chief . . .' and 'masculine' James Cook, 28 Apr 1769, Journal 85, southseas.nla.gov.au/journals/cook/17690428.html; 'lusty' Joseph Banks,

28 Apr 1769, southseas.nla.gov.au/journals/banks/17690428.html; Bougainville's language in describing Tahitian women, van Tilburg 2006; 'A fine and . . .' Bougainville, 5–6 Apr 1768, Dunmore 2002b p59; 'In spite of . . .' Bougainville, Forster 1772 pp218–19. **p160** 'The king is . . .' Caro, 8–9 Apr 1768, Dunmore 2002b p206; wealth redistribution in Māori culture, Maning 1863 Ch7, also Sahlins 1974 Ch4. **p161** 'What is there . . .' Commerson, 'Sommaire d'observations d'histoire naturelle', 24 Dec 1766, Taillemite 1977 p514. **p162** Birdnose wrasse *Gomphosus varius* or *Gomphosus bleu,* Lacépède 1801 V3 pp101–103, and specimens collected by Commerson and Barret, Muséum National d'Histoire Naturelle, MNHN-IC-8240–8241. **p163** 'The act of . . .' and 'Every stranger is . . .' Commerson 1769, Lansdown 2006 p81; 'I was strolling . . .' Nassau-Siegen, Dunmore 2002b p284. **p164** 'the most effective . . .' Orwell 1948 p90; interpretations of the sexuality of Tahitian women, McAlpin 2017; Elliston 1999; and particularly Hermes 2009, who discusses the hazards of Western interpretations. **p165** 'He had hardly . . .' Bougainville, Forster 1772 p219. **pp166–167** The descriptions of Ahutoru aboard the *Étoile,* Caro and Vivez, Dunmore 2002b pp204, 226–29. **pp167–168** Negotiations with Reti (or Ereti), Bougainville, 7 Apr 1766, Dunmore 2002b p62; Jeanne's encounter with the Tahitians, Vivez, Caro and Bougainville, Dunmore 2002b pp229, 205, 97. **p169** *Māhū* is mentioned in Vivez, Dunmore 2002b p229; '*Māhū,* that can . . .' Elliston 1999. **p170** 'a sense of . . .' Condamine, Harrison 2012 p44. **p171** 'M. de Commerson . . .' Bougainville, 11 Apr 1768, Dunmore 2002b p68; shells collected in Tahiti on the Bougainville voyage, Martens 1872.

11. Confession

p177 'his only thought . . .' Bougainville, 21–22 Apr 1768, Dunmore 2002b p77. **p178** Commerson and Bougainville's meeting, Bougainville, 21–22 Apr 1768, Dunmore 2002b p77; Fesche and Commerson, 18–20 Apr 1768, Taillemate 1977 pp96, 474. **p180** Ahutoru's attitude to Vanuatuans, Duclos-Guyot, 7–8 May 1768, Taillemite 1977 p476; encounters in Vanuatu, Caro and Bougavinille, Dunmore 2002b pp212, 93. **p183** 'its most dreaded . . .' Melville 1892 p262. **pp185–186** 'It had been . . .' Saint-Germain, Monnier 1993 p93. **p188** Jeanne's interview revealing her identity, Bougainville 1771 pp253–54; and Forster 1772 pp300–301.

12. In the Teeth of the Reef

p189 'alarming her decency' Bougainville, Dunmore 2002b p97 **p192** 'suffer any unpleasantness' Bougainville, Dunmore 2002b p97; breast-binding, Vivez, Versailles manuscript, Dunmore 2002b p230 and in more detail in the later Rochefort version, Dunmore 2002a p124; also Saint-Germain, Monnier 1993 p93; Bougainville, Commerson and Caro all described birds rising from the sandbank, with Caro naming it Gannet Island. The birds listed are known to have bred here historically. **p199** Australia's inland sea, Clode 2015. **p202** Early description of Australia (New Holland) as dry and sandy, William Dampier, 1688, Flannery 1998 p26. **p203** 'We are sailing . . .' Caro, Dunmore 2002b p213. **p204** 'The latter disgusting . . .' Lesson, Glaubrecht and Podlacha 2010. **pp205–206** 'He was killed . . .', 'We found it . . .' and 'with great gusto' Saint-Germain, Dunmore 2002a p121. Bougainville and Nassau-Siegen also comment positively on eating rats, which sold for 12 sols, Vivez, Dunmore 2002b p242; 'We had to . . .' Vivez, Versailles version, Taillemite 1977 V2 p260.

13. Nature's Beauty and Terror

p209 'Our crews are . . .' Bougainville, 18–19 Jul 1768, Dunmore 2002b p123; further descriptions of Port Praslin, Bougainville, 7 Jul 1768, Dunmore 2002b p118. **p210** 'Art would struggle . . .' Bougainville, 14 Jul 1768, Dunmore 2002b p122; 'The reefs that . . .' Bougainville, Dunmore 2002b p123. **p212** Zebra moray eel (*Echidna zebra)* specimen, Muséum National d'Histoire Naturelle, MNHN IC B2376, coldb.mnhn.fr/catalognumber/mnhn/ic/b-2376; plants from specimens collected by Commerson in Papua New Guinea registered in the Muséum National d'Histoire Naturelle vascular plant catalogue science.mnhn.fr/institution/mnhn/collection/p/item/list?countryCode=PG&recordedBy=Commerson; 'The name of . . .' Specimen label for *Boea magellanica*, G00300724, Phanerogamic Herbarium, Conservatoire et Jardin Botaniques, Geneva. **p214** 'It has been . . .' Duclos-Guyot, Taillemite

1977 V2 p485. **p216** Further details of Veron's astronomical work and achievements, Gascoigne 2015. **p217** 'These 24 hours . . .' Bougainville's journal, 22 Jul 1768, Dunmore 2002b p125 (the earthquake is not mentioned in the official narrative); 'It has rained . . .' Caro, 22 Jul 1768, Taillemite 1977 V2 p346. **p218** 'The officer reported . . .' Fesche, 21–22 Jul 1768, Taillemite 1977 V2 p120; 'On the 21st . . .' Nassau-Siegen, 21 Jul 1768 (section deleted from both later versions), Taillemite 1977 V2 **p408. pp218–219** 'If the reader . . .' Vivez, Versailles manuscript, with deleted sections from the earlier Rochefort version, Taillemite 1977 V2 pp267–68. *Concha veneris* or the shell of Venus often refers to cowries, a common symbol of female sexuality across cultures. The Venus clam *Pitar dione* was also called *Concha veneris* ('Chama', Society of Gentlemen 1763 V1 p537) and was infamously described by Linneaus as resembling a vulva; 'The weather is . . .' Duclos-Guyot, 18 Jul 1768, Taillemite 1977 V2 p485. **p220** 'evaded ambush by . . .' Commerson, Cap 1861 p37, 'If by chance . . .' Commerson, Crassous 1799 p160; 'Historians of the . . .' Ridley 2011 p189. **p221** Statistics on violence against women, UNODC 2018; lack of stories about women travelling alone and adventuring, McNichol 2015 and Veselka 2014.

14. Starvation and Scurvy

p223 'wet and unpleasant . . .' Bougainville, Dunmore 2002b p125; 'their enormous and . . .' Bougainville, 28–29 Jul 1768, Dunmore 2002b p128. **p224** 'What food, Good . . .' Bougainville, Dunmore 2002b p133; eighteenth-century birth control in France, Théré 1999; Doig and Sturzer 2014. **p225** 'There have been . . .' Bougainville, Dunmore 2002b p135; descriptions of Buru (also known as Bouro, Boro, Boeroe and Bouru), Bougainville, Dunmore 2002b pp144–46. **p226** Plants taken from specimens collected by Commerson in Buru (Indonesia) registered in the Muséum National d'Histoire Naturelle vascular plant catalogue, science.mnhn.fr/institution/mnhn/collection/p/item/list?countryCode=PG&-recordedBy=Commerson; birds collected by Commerson at Buru, Buffon and Cuvier 1831–32 V24 p315, V25 p143. Only one bee-eater is found on Buru but seven kingfishers (two endemic), avibase.bsc-eoc.org/checklist.jsp?region=IDmabu; Biographical details of Sonnerat, Ly-Tio-Faine 2008. **p227** Parrot named after Barret (spelt in Latin as Baret), Monnier 1993 pp99–100. There are twelve parrots recorded from Buru, six of which are endemic, avibase.bsc-eoc.org/checklist.jsp?region=IDmabu. **p228** Approach to Batavia, Bougainville, Forster 1772 p398. **pp229–230** Albertus Seba's illustrated collection, Seba 1734–65; current number of cone shells, Puillandre *et al.* 2015; 'We had scarce . . .' Bougainville, Forster 1772 p446; account of their stay in Batavia, Bougainville, Dunmore 2002b p170; Ahutoru's illness in Batavia, Anon. 1772 p51. **p230** Byron lost nine of his 285 crew, mostly to malaria, with no deaths from scurvy. Wallis lost three men from scurvy from a crew of about 160 and 31 of 89 crew to scurvy on his poorly equipped consort ship, Williams 2013 and Wallis 1965. **p233** Human dispersal in South America, Borrero 1999. **p234** 'It is easy . . .' Commerson to Beau, 30 Nov 1768, Morel 2012a p3; 'Of what use . . .' Saint-Germain, Dunmore 2002a p132.

15. Journey's End

p236 'What a fate . . .' Bougainville, Dunmore 2002b p178; sections of this chapter have been adapted from Clode 2017. **p238** 'The town has . . .' Bernardin de Saint Pierre 1775 pp54–55; arrival in Port Louis, Mauritius, Bougainville, Dunmore 2002b p178. **p239** 'I left behind . . .' Bougainville, Dunmore 2002b p179; Poivre is described by Bernardin de Saint-Pierre in M. Souriau 1901. 'Une aventure de Bernardin de Saint-Pierre à l'Ile de France' p397, Morel 2010 p7. **p240** 'The Court, I . . .' Bougainville, 28 May 1768, Dunmore 2002b p97. **p246** 'He went out . . .' Bougainville, Forster 1772 pp265–56; Ahutoru's time in Paris, Anon. 1772; 'Were these Savages?' Fesche, Dunmore 2002b p266. **p247** Commerson's quotes about Tahiti, Commerson 1769; 'I am a . . .' Bougainville, Forster 1772 p8. **p248** Catholic Parish registers can be found at Cercle de Généalogie Maurice, Rodrigues, www.cgmrgenealogie.org/actes/chercher.php. Glynis Ridley suggests that a man named Bonnefoy, who was Flinders' interpreter when Flinders was imprisoned at Bezac's estate in Flacq in Mauritius, was Jeanne's missing child, born in 1768. The interpreter Bonnefoy was appointed by General Decaën in 1803 and was mentioned by Flinders, not in Flacq but at the Garden Prison in Port Louis

(Flinders 1814a p896). Bonnefoy was a common name on Mauritius and there is no other reason to connect the interpreter to Jeanne Barret. **p249** Jeanne is described as living in a large house with Commerson next door to Poivre, Lieberman 2006 p84; 'Poivre, who was . . .' Beau to Minister Turgot, 1774, Commerson's dossier, Archives Nationales (AN Col E89), transcribed at www.pierre-poivre.fr/doc-74-an-c.pdfs; Bernardin de Saint-Pierre's description of the gardens around Mon Plaisir, including 'a multitude of . . .' and 'A brook circulates . . .' 'Une aventure de Bernardin de Saint-Pierre à l'Ile de France' p400, Morel 2010 p8. **p250** The story of the *bonnet carré*, Lesson 1839 p52. **p251** 'it was difficult . . .' Lalande in a letter to the Royal Academy of Sciences, François Rozier, 1776 'Observations on physics, natural history and the arts', 8.2: 357–63 reproduced at www.pierre-poivre.fr/doc-76–11-mois.pdf; 'I hardly ever . . .' Françoise Robin (Mme Poivre) to Bernardin de Saint-Pierre (Le Havre 152–22, BSP.131) quoted in Morel 2010 p20. **p252** Poivre and Commerson's residence in the Hôtel de l'Intendance, Morel 2011 p4. **p253** A detailed inventory of Commerson's archival material in the library of the Muséum National d'Histoire Naturelle, Laissus 1978; Sonnerat had arrived in Mauritius at the same time as Jeanne and Commerson, and illustrated for Commerson until October 1770, when he sailed to the Moluccas with his uncle to obtain spices. He continued with his work on his return in June 1772; 'who make me . . .' Commerson to Bernard de Jussieu, 6 Feb 1770, Cap 1861 p122. **p254** 'the daughter of . . .' Françoise Robin (Mme Poivre) to Bernardin de Saint-Pierre (Le Havre 152–22, BSP.131), Morel 2010 p20; 'It is a . . .' Commerson to Joseph-François Charpentier de Cossigny, 19 Apr 1770, Cap 1861 p129. **p255** Details of *Baretia bonafida* (now *Turraea rutilans*), Bosser *et al.* 1997 pp4, 6; Commerson's Latin description of *Baretia bonafida*, Cap 1861 p37; 'very doubtful sexual . . .' This often-quoted statement was not made by Commerson but by Cap 1861 fn18 p37; frequency of plant hermaphroditism, Barret Spencer and Hough 2013.

16. A Woman of Means

pp257–258 Concession of land to Jeanne Barret, signed by Pierre Poivre, 12 Aug 1770, 'Nanette Anne-Baret Jeanne', Mauritius National Archives LC 7 pp75–76. **p259** Photos of early Port Louis streetscapes, Vintage Mauritius, vintagemauritius.org/port-louis/port-louis-arsenal-street-and-the-citadel-early-1920s. **p260** Details of the sexuality of molluscs, Collin 2013. **p261** 'I embarked from . . .' Commerson, 'Voyage de Madagascar in 1770', Muséum National d'Histoire Naturelle archives, MS 887 II, YL 44, 'Philbert Commerson, voyage à Madagascar en 1770', www.pierre-poivre.fr/doc-70–10–11.pdf. **p262** Commerson says three and a half months but Morel calculates that it cannot have been more than two, Morel 2012a p9; 'I do not . . .' Commerson to Lalande, Morel 2012b p13. **p263** The story of Ahutoru and the mulberry tree, Bernardin de Saint-Pierre, 1810; Ahutoru's stay in Mauritius, Poivre to Bertin, Minister of State, 3 Nov 1770, La Harpe 1780 pp199–202; 'took pleasure in . . .' Vivez, Rochefort manuscript, Taillemite 1977 V2 pp139–40; population of Port Louis, Griffiths *et al.* 2018. **p264** 'open, gay and . . .' and 'the hour of . . .', Bernardin de Saint-Pierre 1775 p184; 'He had acquired . . .' Rochon 'Mission for the return of Aoutourou to Tahiti' (1771–1772)' pp3–4, at www.pierre-poivre.fr/doc-71-an.pdf; 'Ahutoru seemed chagrined . . .' Bernardin de Saint-Pierre 1775 p85. **pp265–266** 'It is a . . .' Commerson to Cossigny, 29 Sep 1770, Cap 1861 pp139–40; acclimatisation and Mauritius, Cheke and Hume 2010 p95. **p267** Commerson's ill health, Commerson's last letter to de Boynes (Secretary of State at the Navy), 17 Oct 1772, Cap 1861 pp169–71; biographical details of Maillart du Mesle, Muteau and Garnier 1859 p185; Amanton 1835; Commerson's precarious financial state, Beau to Minister Turgot, 1774, Commerson's dossier, Archives Nationales (AN Col E89), www.pierre-poivre.fr/doc-74-an-c.pdf; Jossigny's relationship with Commerson, Bour 2015 p421; Maillart-Dumesle, 6 Oct 1772, Archives Nationales (AN ColE/231, 477), at www.pierre-poivre.fr/doc-72–10–6b.pdf; 'I found myself . . .' Commerson to Lemonnier, 27 Oct 1772, Municipal Library of Nantes (MS2423), www.pierre-poivre.fr/doc-72–10–27.pdf. This is Commerson's last known letter. **pp267–268** 'He did not . . .' Maillart to the Minister, 15 Mar 1773, Commerson Dossier, Archives Nationales (AN ColE. 89 vue 573), www.pierre-poivre.fr/doc-73–3–15.pdf. Details of Commerson's address from Gilles Pacaud '7–19 P. Commerson achéte une maison au 194 rue des Pamplemousses à Port-Louis (Île de France)' www.jeanne-barret-tourdumonde.fr. **pp268–269** 'Each man is . . .' Bernardin de Saint-Pierre 1775 p92; 'When he mentioned . . .' and 'in a retired . . .' Lalande 1775, *The Monthly Review* version p600. **p269** Listed as Jęan Nicolas Bezacq in 'Recensement de la population a l'Ile de France', 28 Nov 1776, MS2282, Mauritius collection 1737–1969, Item 13, National Library of Australia. Also Jean Besac in Macquet 1887 p54; 'He had a . . .' Montessus 1889 pp252–53. **p271** 'This plant showing . . .' Commerson, Cap 1861 p37. This Latin translation is open to some interpretation but past published transcriptions and translations are often quite poor and completely skip key phrases – particularly about Jeanne being unharmed by threats to her life and virtue, as well as the ambiguous last line about her detractors (quoted on p220). **p272** 'She will be . . .' Commerson, Cap 1861 p37.

17. Unexpected Treasures

p273 'to put in . . .' Commerson 1774 p13. No women's clothes were found in Commerson's Port Louis apartment after his death. **p274** Details of inn-keeping and fine for serving alcohol, Christinat 1993 p49 and Gilles Pacaud 'La vie de Jeanne Barret á l'île de France (Maurice) 1768–1774' (particularly 7–10 to 7–12), www.jeanne-barret-tourdumonde.fr. **pp274–275** Transcriptions of the inventory of Commerson's Port Louis apartment by Daniel Margottat can be found in section 13 of www.jeanne-barret-tourdumonde.fr. The inventory of Commerson's collections shipped to France was attached to a letter from Maillart to the Minister, 9 Nov 1773, Archives Nationales (AN Col C/4/34, folio 196), www.pierre-poivre.fr/doc-74-6-3.pdf. There are many possible translations for cajou, including kagu or cagou (*Rynochetos jubatus)* an endemic bird from New Caledonia first described by Jules Verreaux in 1860, or cagou, which in older dictionaries refers to an owl (e.g. Abel Boyer, *The Royal Dictionary Abridged: In Two Parts: I. French and English, II. English and French*. 13th edition (London, Bathurst *et al.* 1771). **p277** 'Jossigny, the draftsman . . .' Bour 2015 p422. **pp277–278** 'which were deposited . . .' Lalande 1775, *The Monthly Review* version p606. **p278** Letter from Beau to Minister Turgot, Monnier 1993 pp173–80 (from the Archives Nationales AN Col. E 89); size of plant collections, Williams 2003 p11. **pp278–279** 'However, we must . . .' Antoine-Laurent de Jusseiu to Archambeau Commerson, 13 Mar 1789, Montessus 1889 p200; International Plant Names Index is at www.ipni.org, with details of Commerson's contributions at www.ipni.org/a/1753-1; re Commerson's stature, a reviewer for *Nature* noted that 'had Commerson lived, he would have left a name second only to that of Linnaeus among eighteenth-century naturalists' (JSG 1909 p430); 'It is astounding . . .' Cuvier, Carrapiço 2018 p261; Commerson's work on reptiles, Bour 1982; on whales, Harmer 1922; Commerson's original notes were published in Buffon and Cuvier 1831–32; Buffon 1799; Lacépède 1801 and 1804; by Adrien and Antoine-Laurent de Jussieu in Saint-Hilaire *et al.* 1829–32; and Lamarck and Jussieu 1803. Details of further information on Mascarene birds and mammals derived from Commerson and other historical material, Hume and Prys-Jones 2005; historic bird specimens in the Muséum National d'Histoire Naturelle, Jouanin 1962. **p281** Essays on Museum of Victoria, Clode 2006. **pp282–283** Inventory of collections in the Paris apartment, Monnier 1993 pp152, 172 from a document in the Archives Nationales, Étude Regnault, ET/84/537. **p283** Size of collections, Commerson to Lalande, 18 April 1771, published in the appendix of a French edition of Cook's voyage, Commerson 1773 p172–3. **pp285–286** 'objects of curiosity . . .' and 'They are never . . .' Commerson, 'Summary of natural history observations' to Praslin, 24 Dec 1766, Taillemite 1977 p516. **p286** 'If ever the . . .' Françoise Robin (Mme Poivre) to Bernardin de Saint-Pierre (Le Havre, 152–16, BSP129), Morel 2010 p22; shell merchants in Paris, Burkhardt 1977 p13 and Dietz 2006; sale of part of the collection to benefit Commerson's son, Dussourd 1987 p74. **pp286–287** Details of shells collected from various Bougainville voyages, and the collections they ended up in, Martens 1872; Dezallier d'Argenville's interests in shells from 'Dezallier d' Argenville's "La Conchyliologie" – A delight for your eyes and soul', *Where the Art Is*, rareprintsworld.wordpress.com/dezallier-d-argenvilles-la-conchyliologie-a-delight-for-your-eyes-and-soul; 'The coasts of . . .' Dezallier d'Argenville 1780 pp120–21; giant helmet shell is illustrated in Dezallier d'Argenville 1780 Plates XXV–XXVII. **p288** Silence of molluscs, Vermeij 2010. **pp288–289** Acte de Mort, Leonard Lanoiselée, 9 May 1773, Archives Municipales Saône-et-Loire, Rosiéres (Toulon-sur-Arroux), Baptêmes,

Mariages, Sépultures 1748–1780, E dépôt 6173–6175, p150, www.archives71.fr/ark:/60535/s00513996d55986e/5139b293acf60; marriage certificate of Jeanne Barret and Jean Dubernat, Archives of Mauritius, Port-Louis, K. A. 61/D, folio 21, p75. Details of Jeanne's wealth are from the marriage contract between Jeanne Barret and Jean Dubernat, 13 May 1774, Mauritius National Archives, Notary Gombault, NA 18/8 13.5.1774; biographical information on Jean Dubernat, Maguet and Miquel 2019; La Giraudais's involvement in unsuccessful trading endeavours, see document no.4 – a memoir of Jean-Joseph Amat, Naval Agent, of his time on Îsle de France 1768–76, www.pierre-poivre.fr/doc-nodate-33.pdf; **p291** Saint-Louis Chapel was built by Mahé de la Bourdonnais in 1737 on Royal Street, Port Louis. The current Saint-Louis Church was built in 1756 but was not completed until 1782, when religious services and many of the interior features and tomb were transferred from the chapel, Vintage Mauritius, vintagemauritius.org/port-louis/port-louis-saint-louis-cathedral-1900s. The witnesses to Jeanne's marriage mentioned on the marriage certificate were Jacques Ronqué, Jacques Lafont, François Chevalier (a 27-year-old *piqueur*, or procurer, for the King's works), Cristophe Sales (50-year-old merchant from southern France) and Charles Ouvrard (a 42-year-old carpenter). The land concession to Jean Dubernat, 5 Jun 1774, Mauritius National Archives Administrateurs Généraux: Port Louis mainly, LC 14, pp40–41 with details of Mauritian land concessions from Macquet 1887. **pp292–293** Details of Aimé Bonnefoy from the marriage contract between Jeanne Barret and Jean Dubernat, 13 May 1774, Mauritius National Archives, Notary Gombault, NA 18/8 13.5.1774; details of Commerson's secret bequests and the unknown child are from unpublished sections of his will and legal documents separate from the official inventory of his Paris apartment, transcribed by Daniel Margottat '13–5 Documents de la succession de P. Commerson. Etude Regnault. 1774' www.jeanne-barret-tourdumonde.fr. This document has led to suggestions that Commerson was paying for his and Jeanne's child to be cared for by Mme Mercedier (who is a beneficiary of Commerson's will and often confused with Jeanne in Commerson's family histories) but I could find no evidence of a child matching this description in the archives of Commerson's home town.

18. Homecoming

p298 Story of Susan Sibbald, Sibbald and Paget Hett 2007; story of Edith Coleman, Clode 2018. **p300** 'Sieur Dubernat with . . .' extract of the passenger list of the *Sympathie* arriving in Bordeaux on 26 Aug 1775, Maguet and Miquel 2019 p4; the timing of their departure is suggested by a report to the notary on 10 Apr 1805, Maguet and Miquel 2019 p5. **p301** Weather details, Marusek 2010 p266; description of the house, Acte d'achat (Dordogne Archives, AD33 – P. Brun, Sainte-Foy-la-Grande, 1775, 3 E 42616, folio 172) and the neighbours from the 1790 Civil Census for Sainte-Foy-la-Grande, Maguet and Miquel 2019 p7. **p302** The bibliography of Jeanne Barret, Maguet *et al.* 2014. **p304** Details of the land transaction, Acte de vente (Dordogne Archives, AD 24—P. B. des Laurens, 1775–76, 3 E 6994, folio 2), Maguet and Miquel 2019 p8; 'The goods are . . .' Accord Jean Dubernat-Jeanne Barret, 10 Apr 1805 (Dordogne Archives, AD 33 – Notaire P. Brun, Sainte-Foy-la-Grande, 3 E 42645, An XIII-An XIV, folio 71), Maguet and Miquel 2019 p6; details of Commerson's estate, 3 Apr 1776, Archives Nationale, MC/ET/LXXXIV/537, Notary Regnault, folio 31; Monnier 1997 pp45, 98. **p306** Inventory of Jeanne's possessions, Miquel 2017. **p307** Gesse or chickling vetch is *Lathyrus sativus*; 'The voyages of . . .' Commerson, Lesson 1839 p7. **pp308–309** 'Jeanne Barré, by . . .' government order awarding Jeanne's pension, 13 Nov 1785, Archives of the Ministére de la Marine, C7 17, Bibliothéque Centrale de la Marine (Vincennes). It's not entirely certain, though, that this money was forthcoming. In 1794 Jeanne granted some Parisian notaries power of attorney to act on her behalf to retrieve arrears owed to her by the Republic (from Dordogne Archives, Contrôle des actes notaires, Sainte-Foy-la-Grande. AD 33–24 Thermidor Year II-17 Ventose Year III (11 Aug 1794 – 7 Mar 1795) 3Q, Maguet and Miquel 2019 p17). **p310** Colette 1929/1968; 'finished her days . . .' Cap 1861 p37. **p311** Jeanne's wills, Miquel 2017.

Epilogue

p314 'she had no . . .' Eliot 1871/1985 p896.

References

Albert, Simon, Javier X. Leon, Alistair R. Grinham *et al.* 2016. 'Interactions between sea-level rise and wave exposure on reef island dynamics in the Solomon Islands', *Environmental Research Letters*, 11: 054011.

Amanton, Claude-Nicolas. 1835. *Galerie Auxonnaise: Biographies de Maillard de Mesle* (X. T. Saunié: Auxonne).

Anon. 1772. 'Characters: An account of a native of Taiti (an island in the South Seas), who accompanied M. de Bougainville to France, in the year 1769)', *Annual Register or a view of the History, Politics and Literature for the Year 1772*, V15 (J. Dodsley: London): 49–52.

——. 1792. *Journal d'un voyage de Genève à Paris par la diligence, fait et 1791* (J. E. Didier: Paris), Bibliothèque nationale de France, Département philosophie, histoire, sciences de l'homme, 8-L29–9, gallica.bnf.fr/ark:/12148/bpt6k102924q.

——. 1806. *Navigazioni di Cook pel grande oceano et intorno al globo*, V2 (Sonzogno e Co: Milan).

Bachaumont, Louis Petit de. 1784. *Mémoires Secrets pour servir à l'histoire de la République des Lettres en France, depuis MDCCLXII jusqu'a nos jours* (John Adamson: London).

Bareille, Patrick. 2019. 'The Morvan diau Pat: Vie et traditions du Morvan et du Nivernais entre le XVIIe et le XXe siècle', lemorvandiaupat.free.fr/index.html.

Barrett Spencer, Charles H. and Josh Hough. 2013. 'Sexual dimorphism in flowering plants', *Journal of Experimental Botany*, 64.1: 67–82.

Bassett, Marnie. 1962. *Realms and Islands: The world voyage of Rose de Freycinet in the corvette* Uranie *1817–1820* (Oxford University Press: London).

Bernardin de Saint-Pierre, Jacques Henri. 1775. *A Voyage to the Island of Mauritius, the Isle of Bourbon, the Cape of Good Hope, with Observations and Reflections upon Nature* (W. Griffin: London).

——. 1810. 'The love of our native land', *New Magazine of Choice Pieces*, V1 (John Perry: London).

——. 1823. 'Harmonies de la nature' in *Œuvres complétes de Jacques-Henri-Bernardin de Saint-Pierre* (Aimé André: Paris).

Blanvillain, Caroline, Chevallier Florent, and Vincent Thenot. 2002. 'Land birds of Tuamotu Archipelago, Polynesia: relative abundance and changes during the 20th century with particular reference to the critically endangered Polynesian ground-dove (*Gallicolumba erythroptera*)', *Biological Conservation*, 103: 139–149.

Bloomfield, Noelene. 2012. *Almost a French Australia: French-British rivalry in the southern oceans* (Halstead Press: Canberra).

Bonnot. 1787. 'Observations faites dans le département des hôpitaux civils', *Journal de Medecine, Chirurgie, Pharmacie etc.*, 72: 387–399.

Borrero, Luis Alberto. 1999. 'The prehistoric exploration and colonisation of Fuego-Patagonia', *Journal of World Prehistory*, 13: 321–355.

Bosser, Jean Marie, Thérésien Cadet, Joseph Guého, and W. Marais. 1997. *Flore des Mascareignes: La Réunion, Maurice, Rodrigues* (Sugar Industry Research Institute: Mauritius).

Bougainville, Louis. A. 1771. *Voyage autour du monde par la frégate du Roi* la Boudeuse *et la flûte* l'Étoile; *en 1788, 1767, 1768 & 1769* (Saillant & Nyon: Paris).

Bour, Roger. 1982. '*Pelomedusa subrufa* (Lacepéde, 1788), *Pelusios subniger* (Lacepéde, 1788) (Reptilia, Chelonii) et le séjour de Philibert Commerson à Madagascar', *Bulletin du Muséum National d'Histoire Naturelle, Paris*, 4: 531–539.

——. 2015. 'Paul Philippe Sanguin de Jossigny (1750–1827), artiste de Philibert Commerson: Les dessins de reptiles de Madagascar, de Rodrigues et des Seychelles', *Zoosystema*, 37: 415–448.

Brown, Nancy M. 2007. *The Far Traveler: Voyages of a Viking Woman* (Harcourt: New York).

Buffon, Georges L. L. 1799. *Histoire Naturelle des Oiseaux*, V6 (l'Imprimerie Royale: Paris).

—— and Frédéric Cuvier. 1831–32. *Oeuvres Complètes de Buffon* (F. D. Pillot: Paris).

Burkhardt, Richard W. 1977. *The Spirit of System: Lamarck and evolutionary biology* (Harvard University Press: Cambridge, Massachusetts).

Cap, Paul-Antoine. 1861. *Philibert Commerson: Naturaliste Voyageur – Étude biographique suivie d'un appendice* (Victor Masson et fils: Paris).

Carrapiço, Francisco. 2018. 'Azolla and Bougainville's *Voyage around the World*' in Fernández, Helena (ed) *Current Advances in Fern Research*, (Springer: Cham).

Cheke, Anthony, S., and Julian P. Hume. 2010. *Lost Land of the Dodo: The ecological history of Mauritius, Réunion and Rodrigues* (Bloomsbury: London).

——. Miguel Pedrono, Roger Bour, *et al.* 2017. 'Giant tortoises spread to western Indian Ocean islands by sea drift in pre-Holocene times, not by later human agency – response to Wilmé *et al.* (2016a)', *Journal of Biogeography*, 44: 1426–1429.

Chrisafis, Angelique. 2013. 'French nuclear tests "showered vast area of Polynesia with radioactivity"', *Guardian*, 4 July, www.theguardian.com/world/2013/jul/03/french-nuclear-tests-polynesia-declassified.

Christinat, Carole. 1995. 'Une femme globe-trotter avec Bougainville: Jeanne Barret (1740–1807)', *Annales de Bourgogne*, 67: 41–55.

Clode, Danielle. 2002/2011. *Killers in Eden: The story of a rare partnership between men and killer whales* (Museums Victoria: Melbourne).

——. 2006. *Continent of Curiosities: A journey through Australian natural history* (Cambridge University Press: Melbourne).

——. 2007/2018. *Voyages to the South Seas: In search of Terres Australes* (Ligature: Sydney).

——. 2015. *Prehistoric Marine Life in Australia's Inland Sea* (Museums Victoria: Melbourne).

——. 2017. 'Mauritius – paradise regained', *Zoomorphic*, V8, 10 March, zoomorphic.net/2017/03/mauritius-paradise-regained.

——. 2018. *The Wasp and the Orchid: The remarkable life of Australian naturalist Edith Coleman* (Picador: Sydney).

Colette, Sidonie-Gabrielle, 1929/1968. *My Mother's House* (Penguin Books: Harmondsworth).

Collin, Rachel. 2013. 'Phylogenetic patterns and phenotypic plasticity of molluscan sexual systems', *Integrative and Comparative Biology*, 53: 723–735.

Commerson, Philibert. 1769. 'Postscript: on the island of New Cythera or Tahiti', *Mercure de France* as quoted in Lansdown, Richard. 2006. *Strangers in the South Seas: The idea of the Pacific in Western thought* (University of Hawai'i Press: Honolulu).

——. 1774. 'Testament singulier de M. Commerson, Docteur en médecine, médecine botaniste and naturaliste du Roi fait 14 and 15th Decembre 1766.' (Unknown publisher: Paris).

——. 1773. 'Lettre de M. Commerson á M. de la Lande, de l'isle de Bourbon, le 18 avril 1771' in Anne François Joachim Fréville, [translator] *Supplément au Voyage de M. de Bougainville, ou Journal d'un Voyage autour du monde fait par MM. Banks et Solander, Aglois, en 1768, 1769, 1770, 1771* (Societe Typographique: Neuchatel).

Covington, Syms. 1839. 'The Journal of Syms Covington, Assistant to Charles Darwin Esq. on the second voyage of the *HMS Beagle* December 1831 – September 1836'. State Library of NSW Manuscripts, MLMSS 2009/Box108/Items5–6, web.archive.org/web/20040406223512/http://austehc.unimelb.edu.au/bsparcs/covingto/contents.htm#contents.

Crassous, Paulin. 1799. 'Lettres de Commerson, contenant un détail succinct de son voyage ouretour du globe, et précédres d'une notice de sa vie, de son charactére et de ses ouvrages' *L'Esprit des Journaux Francais et Étrangers* 8:151–184.

Creighton, Margaret S., and Lisa Norling. 1996. *Iron Men, Wooden Women: Gender and seafaring in the Atlantic World, 1700–1920* (John Hopkins University Press: Baltimore).

Crestey, Nicole. 2011. 'L'Affaire Jeanne Barret' in Marie-Françoise Bosquet, and Chantale Meure (eds), *Le Féminin en Orient et en Occident, du Moyen Âge à nos jours: Mythes et réalités* (Presses de l'Université de Saint-Étienne: Saint-Étienne): 327–343.

Darwin, Charles. 1887. *The Life and Letters of Charles Darwin including an Autobiographical Chapter*, V1–3 (William Clowes & Sons: London), darwin-online.org.uk/EditorialIntroductions/Freeman_LifeandLettersandAutobiography.html.

Delrieu, André. 1831. *Les Enfants-Trouvés* (Ladcovat: Paris).
Demerliac, Alain. 1995. *Nomenclature des Navires Français de 1715–1774* (Ancre: Nice).
Dening, Greg. 1997. 'Empowering imaginations', *The Contemporary Pacific*, 9.2: 419–429.
Dezallier d'Argenville, Antoine-Joseph. 1780. *La Conchyliologie*. Third edition (Guillaume de Bure: Paris).
Diderot, Denis. 1796. *Supplément au Voyage de Bougainville* (Chevet: Paris).
Dietz, Bettina. 2006. 'Mobile objects: The space of shells in eighteenth-century France', *British Journal for the History of Science*, 39: 363–382.
Doig, Ann Kathleen, and Felicia B. Sturzer. 2014. *Women, Gender and Disease in Eighteenth-Century England and France* (Cambridge Scholars Publishing: Newcastle-upon-Tyne).
Downing, Karen. 2014. *Restless Men: Masculinity and Robinson Crusoe 1788–1840* (Palgrave Macmillan: Basingstoke).
Drake, Theodore G. H. 1940. 'The wet nurse in France in the eighteenth century', *Bulletin of the History of Medicine*, 8: 934–948.
Druett, Joan. 2000. *She Captains: Heroines and hellions of the sea* (Simon and Schuster: New York).
Duffin, Christopher. 2008. 'Fossils and folklore', *Ethical Record*, 113: 17–21.
Dunmore, John. 2002a. *Monsieur Baret: First woman around the world 1766–1768* (Heritage Press: Auckland).
——. ed. 2002b. *The Pacific Journal of Louis-Antoine de Bougainville: 1767–1768* (Hakluyt Society: London).
——. 2015, *Storms and Dreams: Louis de Bougainville: Soldier, explorer, statesman* (Exisle Publishing: Auckland).
Durand, Frederique, and Julian Wiethold. 2014. 'Social status and plant food diet in Bibracte, Morvan (Burgundy, France)', in Alexandre Chevalier, Elena Marinova, and Leonor Peña-Chocarro. eds. *Plants and People: Choices and diversity through time* (Oxbow Books: Oxford): 412–420, 454–466.
Dussourd, Henriette. 1968. *Toulon-sur-Arroux* (Pottier et Cie: Moulins).
——. 1987. *Jeanne Baret (1740–1816): Première femme autour du monde* (Pottier: Moulins).
Duvat, Virginie K. E., Bernard Salvat, and Camille Salmon. 2017. 'Drivers of shoreline change in atoll reef islands of the Tuamotu Archipelago, French Polynesia', *Global and Planetary Change*, 158: 134–154.
Duyker, Edward. 2003. *Citizen Labillardière: A naturalist's life in revolution and exploration 1755–1834* (Miegunyah Press: Melbourne).
—— and Maryse Duyker. 2006. *Bruny d'Entrecasteaux: Voyage to Australia and the Pacific 1791–1793* (Miegunyah Press: Melbourne).
Eckstein, Lars, and Anja Schwarz. 2018. 'The making of Tupaia's map: a story of the extent and mastery of Polynesian navigation, competing systems of wayfinding on James Cook's *Endeavour*, and the invention of an ingenious cartographic system', *Journal of Pacific History*, 54: 1–95.
Édouard, Victor. 1974. 'Le docteur Philibert Commerson, compagnon de Bougainville, et son valet', *Visage de L'Ain*, 132: 19–28.
Eliot, George (Mary Ann Evans), 1871/1985. *Middlemarch* (Penguin Classics: Harmondsworth).
Elliston, Deborah A. 1999. 'Negotiating transnational sexual economies: female māhū and same-sex sexuality in "Tahiti and her islands"', in Evelyn Blackwood, and Saskia Wieringa (eds). *Female Desires: Same-sex relations and transgender practices across cultures* (Columbia University Press: New York): 232–252.
Farge, Arlette. 1993. *Fragile Lives: Violence, power and solidarity in eighteenth-century Paris* (Polity Press: Cambridge).
Finney, Ben R. 1979. *Hokule'a: The way to Tahiti* (Mead Dodd: New York).
Fitz Roy, Robert (ed.). 1839. *Narrative of the surveying Voyages of His Majesty's ships Adventure and Beagle*, V1–3 (Henry Colburn: London).
Flannery, Tim (ed.). 1998. *The Explorers* (Text: Melbourne).
Flinders, Matthew. 1814a. *A Voyage to Terra Australis*, V2 (G. & W. Nicol: London).
——. 1814b. 'Biographical memoir of Captain Matthew Flinders, RN', *Naval Chronicle for 1814*, V32 (Joyce Gould: London).

Forster, John R. tr. 1772. Louis Bougainville. *A Voyage Round the World. Performed by order of His Most Christian Majesty, in the years 1766, 1767, 1768, and 1769* (Nourse: London).

Gascoigne, John. 2015. 'Navigating the Pacific from Bougainville to Dumont d'Urville: French approaches to determining longitude, 1766–1840', in Richard Dunn, and Rebekah Higgitt (eds). *Navigational Enterprises in Europe and its Empires, 1730–1850* (Palgrave Macmillan: London) 180–197.

Gelbart, Nina R. 2016. 'Adjusting the lens: Locating early modern women of science', *Early Modern Women: An interdisciplinary journal*, 11: 116–127.

Ghabut, Marie-Hélène. 1998. 'Female as other: The subversion of the canon through female figures in Diderot's work', *Diderot Studies*, 27: 57–66.

Ghosh, Amitav. 2004. *The Hungry Tide* (Harper Collins: London).

Glaubrecht, Matthais, and Kathrin Podlacha. 2010. 'Freshwater gastropods from early voyages into the Indo-West Pacific: the "melaniids" (Cerithioidea, Thiaridae) from the French *La Coquille* navigation 1822–1825', *Zoosystema Evolution*, 86: 185–211.

Griffiths, Owen L., Jean-Marie Huron, Marina Carter, *et al.* 2018. *From Piastres to Polymer: A history of paper money in Mauritius from 1720 to 2017* (Bioculture Press: Mauritius).

Harmer, Sidney, F. 1922. 'On Commerson's dolphin and other species of *Cephalorhynchus*', *Proceedings of the Zoological Society*, 92: 3 (No. 43): 627–638.

Harrison, Carol. 2012. 'Replotting the ethnographic romance: Revolutionary Frenchmen in the Pacific, 1768–1804', *Journal of the History of Sexuality*, 21: 39–59.

Hau'ofa, Epeli. 2008. 'Pasts to remember', *We Are the Ocean: Selected works* (University of Hawai'i Press: Honolulu).

Helm, Rebecca. 2019. 'How plastic cleanup threatens the oceans living islands', *The Atlantic*, 22 Jan, www.theatlantic.com/science/archive/2019/01/ocean-cleanup-project-could-destroy-neuston/580693.

Henry, Louis, and Yves Blayo. 1975. 'La population de la France de 1740 à 1860', *Population*, 30–1: 71–122.

Hermes, Katherine A. 2009. 'Getting nailed: Re-inventing the European-Pacific encounter in the age of global capital', in Karen A. Ritzenhoff and Katherine A. Hermes. eds. *Sex and Sexuality in a Feminist World* (Cambridge Scholars Publishing: Newcastle upon Tyne): 372–383.

Hugo, Victor. 1831. *Notre-Dame de Paris* (Charles Gosselin Libraire: Paris).

Hume, Julian P, and Robert P. Prys-Jones. 2005. 'New discoveries from old sources, with reference to the original bird and mammal fauna of the Mascarene Islands, Indian Ocean', *Zoologische Mededelingen*, 79: 85–95.

Jenkins, Alan C. 1978. *The Naturalists: Pioneers of natural history* (Hamish Hamilton: London).

Jolinon, Jean-Claude. 2005. 'Jeanne Baret, une femme autour du monde', in Jean-Pierre Changeux. ed. *La Lumière au Siècle des Lumières et aujourd'hui* (Odile Jacob: Paris).

Jones, Elin, F. 2016. 'Masculinity, materiality and space onboard the royal naval ship, 1756–1815', PhD thesis, Queen Mary University of London.

Jouanin, Christian. 1962. 'Inventaire des oiseaux éteints ou en voie d'extinction conservés au Muséum de Paris', *Terre et Vie*, 109: 257–301.

JSG. 1909. 'A great naturalist', *Nature*, 80: 430–431.

Kaplan, Steven L. 1982. 'The famine plot: Persuasion in eighteenth-century France', *Transactions of the American Philosophical Society*, 72: 1–79.

Klem, Jonathon. 2017. 'Shells and stones: A functional examination of the Tuamoutus adze kit', MA thesis, University of Hawai'i at Manoa, scholarspace.manoa.hawaii.edu/handle/10125/62090

Lacépède, Bernard G. 1801. *Histoire Naturelle des Poissons* (Plassan: Paris).

——. 1804. *Histoire Naturelle des Cétacés* (Plassan: Paris).

Lack, H. Walter. 2012. 'The discovery, naming and typification of *Bougainvillea spectabilis* (Nyctaginaceae)', *Willdenowia*, 42: 117–126.

La-Croix, Jean F. 1788. *Dictionnaire Portatif des Femmes Célèbres* (Belin et Volland: Paris).

La Harpe, Jean-François de. 1780. 'IV. M. de Bougainville', *Abrégé de l'Histoire Générale des Voyages continué par Comeiras*, V19 (Hotel de Thou: Paris).

Laissus, Yves. 1978. 'Catalogue des manuscrits de Philibert Commerson (1727–1773) conservés à la Bibliothèque centrale du Muséum national d'Histoire naturelle (Paris)', *Revue d'Histoire des Sciences*, 31.2: 131–162.

Lalande, Jérôme. 1775. 'Éloge de M. Commerson', *Observations sur la Physique, sur l'Histoire naturelle et sur les Arts*, V: 89–120. An English extract translated in 1776 as 'An Account of Commerson and his voyage around the world', *The Monthly Review*, 53: 599–607.

Lamarck, Jean-Baptiste. 1804. *Encyclopédie Méthodique: Botanique* (Agasse: Paris).

——. and Antoine Laurent de Jussieu. 1803. *Histoire Naturelle des Végétaux, classés par Familles*, V1–9 (de l'imprimerie de Crapelet, chez Deterville: Paris).

Lansdown, Richard. 2006. *Strangers in the South Seas: The idea of the Pacific in Western thought* (University of Hawai'i Press: Honolulu).

Le Brun, Dominique. 2019. *Bougainville* (Tallandier: Ponant).

Lesson, René P. 1839. *Voyage autour du monde entrepris par ordre du gouvernement sur la Corvette* La Coquille (Gregoire, Wouters: Bruxelles).

Lieberman, Harry. 2006. *The Travelers' World: Europe to the Pacific* (Harvard University Press: Cambridge, Massachusetts).

Lignereux, Yves. 2004. 'Philibert Commerson, médecin-naturaliste du Roi (1727–1773) ou la traversée inachevée', *Bulletin du Centre d'Étude d'Histoire de la Médecine*, 47: 7–51.

Lodi, Francesca. 1985. 'Dall'Acqua, Giuseppe', *Dizionario Biografico degli Italiani*, V31 (Trecanni: Rome).

London, Jack. 1911. 'The Seed of McCoy', *South Seas Tales* (Macmillan: New York) www.gutenberg.org/files/1208/1208-h/1208-h.

Ly-Tio-Faine, Madeleine. 2008. 'Pierre Sonnerat', in *Complete Dictionary of Scientific Biography* (Charles Scribner's Sons: Detroit).

MacMasters, Rowland. 2015. '"I hate to hear of women on board": Women aboard war ships', *Persuasions On-Line*, 36.

Macquet, Adolphe. 1887. *Précis-terrier de l'île Maurice, ou, Table générale de toutes les concessions faites dans les divers districts de la colonie: avec notes techniques* (Cernéen: Maurice), nla.gov.au/tarkine/nla.obj-1114669434.

Maguet, Nicolle, and Sophie Miquel. 2019. 'De l'océan Indien aux rives de la Dordogne: le retour de Jeanne Barret après son tour du monde. Jeanne Barret et Jean Dubernat, propriétés et familles en Dordogne et en Gironde', *Cahier des Amis de Sainte-Foy*, 114: 15–42.

——. Sophie Miquel, and Françoise Raluy. 2014. 'Bibliographe Jeanne Barret', *Bulletin de la Société Botanique du Périgord*, Special Bulletin 6: 1–12.

Maning, Frederick Edward. 1863. *Old New Zealand: Being incidents of native customs and character* (Smith, Elder and Co.: London).

Martens, Eduard von. 1872. 'Conchylien von Bougainville's Reise', *Malakozoologische Blätter*, 19: 49–65.

Marusek, James A. 2010. 'A chronological listing of early weather events', wattsupwiththat.files.wordpress.com/2011/09/weather1.pdf.

Masefield, John. 1937. *Sea Life in Nelson's Time* (Methuen: London).

Mayr, Ernst. 1982. *The Growth of Biological Thought: Diversity, evolution and inheritance*, (Harvard University Press: Cambridge, Massachusetts).

McAlpin, Mary. 2017. 'Rape in Paradise: Naturalizing sexual violence in Diderot's Tahitian reverie', *Eighteenth-Century Studies*, 50: 289–302.

McBride, Theresa M. 1974. 'Social mobility for the lower classes: Domestic servants in France', *Journal of Social History*, 8: 63–78.

McNichol, Glynis. 2015. 'Women on road trips aren't tragedies waiting to happen. Like men we're free', *Guardian*, 7 Aug, www.theguardian.com/commentisfree/2015/aug/06/women-road-trips-freedom-narratives.

Melville, Herman. 1888. 'The Maldive Shark', *John Marr and Other Sailors* (De Vinne Press: New York).

——. 1892. *Moby Dick; Or, the Whale* (St Botolph Society: Boston).

Mendham, Matthew D. 2015. 'Rousseau's discarded children: The panoply of excuses and the question of hypocrisy', *History of European Ideas*, 41.1: 131–152.

Meyer, Jean-Yves, and Jacques Florence. 1996. 'Tahiti's native flora endangered by the invasion of *Miconia calvescens* DC (Melastomataceae)', *Journal of Biogeography*, 23: 775–81.

Milet-Mureau, Louis Antoine. 1799. *A Voyage round the World performed in the Years 1785, 1786, and 1788 by the* Boussole *and* Astrolabe *under the command of J.F.G. de la Pérouse* V1–3 (A. Hamilton: London).

Miquel, Sophie. 2017. 'Les testaments de Jeanne Barret, première femme à fair le tour de la terre, et de son époux périgordin Jean Dubernat', *Bulletin de la Société Historique et Archéologique du Périgord*, 144: 771–782.

Monnier, Jeannine. 1993. *Philibert Commerson le Découvreur du Bougainvillier* (Association Saint-Guignefort: Châtillon-sur-Chalaronne).

Montessus, Ferdinand B. 1889. 'Martyrologe et biographie de Commerson', *Bulletins de la Société des Sciences Naturelles de Saône-et-Loire*, 3: 78–290.

Morel, Jean-Paul. 2010. 'Bernardin de Saint-Pierre à l'Ile de France', *Sur la Vie de Monsieur Poivre: Une légende revisitée*, www.pierre-poivre.fr/Bernardin-a-Isle-de-France.pdf.

——. 2011. 'Monplaisir, un jardin bien nommé: Deuxième partie: 1767–1772', *Sur la Vie de Monsieur Poivre: une légende revisitée*, www.pierre-poivre.fr/Monplaisir-intro.html.

——. 2012a. 'Eléments biographiques sur Philibert Commerson', *Sur la Vie de Monsieur Poivre: Une légende revisitée*, www.pierre-poivre.fr/doc-nodate-32.pdf.

——. 2012b. 'Philibert Commerson à Madagascar et à Bourbon', *Sur la vie de Monsieur Poivre: Une légende revisitée,* www.pierre-poivre.fr/Commerson-Madagascar-Bourbon.pdf.

Morton, Adam, and Penny Stephens. 2015. 'The vanishing island', Climate for Change Series, *Sydney Morning Herald,* www.smh.com.au/interactive/2015/the-vanishing-island.

Mukerji, Chandra. 2008. 'Women engineers and the culture of the Pyrénées: Indigenous knowledge and engineering in seventeenth-century France', in Pamela Smith, and Benjamin Schmidt (eds). *Knowledge and its Making in the Early Modern World* (Chicago: University of Chicago Press): 19–44.

Muteau, Charles, and Joseph Garnier. 1859. 'Maillart du Mesle (Jacques)', *Galerie Bourguignonne,* V2 (Picard, Lamarch: Dijon).

Neall, Vincent E., and Steven A. Trewick. 2008. 'The age and origin of the Pacific islands: a geological overview', *Philosophical Transactions of the Royal Society of London. Series B, Biological Sciences*, 363: 3293–3308.

Niaux, Roland. 2000. 'Histoire and archéologie en Morvan et Bourgogne', sites.google.com/site/vniaux.

Oliver, Samuel P. 1909. *The Life of Philibert Commerson* (John Murray: London).

Orwell, George. 1948. *1984* (Signet Classics: New York).

Pellegrin, Nicole. 1999. 'Le genre et l'habit. Figures du transvestisme féminin sous l'Ancien Régime', *Clio: Femmes, genre, histoire*, 10.

Peters, Andy. 2013. *Ship Decoration: 1630–1780* (Seaforth Publishing: Barnsley UK).

Pietsch, Roland. 2004. 'Ships' boys and youth culture in eighteenth-century Britain: The navy recruits of the London Marine Society', *The Northern Mariner/Le marin du nord,* XIV.4: 11–24.

Puillandre, Nicolas, Thomas F. Duda, Christopher Meyer, *et al.* 2015. 'One, four or 100 genera? A new classification of the cone snails', *Journal of Molluscan Studies*, 81: 1–23.

Régnier, Claire, Benoit Fontaine, and Philippe Bouchet. 2009. 'Not knowing, not recording, not listing: Numerous unnoticed mollusk extinctions', *Conservation Biology*, 23: 1214–1221.

Ridley, Glynis. 2011. *The Discovery of Jeanne Baret* (Fourth Estate: Sydney).

Robb, Graham. 2007. *The Discovery of France* (Picador: London).

Role, Andre. 1973. 'Vie aventureuse d'un savant: Philibert Commerson, martyr de la botanique (1727–1773)', *Académie de la Réunion*: 151–172.

Root, Hilton L. 1987. *Peasants and King in Burgundy: Agrarian foundations of French absolutism* (University of California Press: Berkeley).

Sahlins, Marshall. 1974. *Stone Age Economics* (Tavistock: London).

Saint-Hilaire, Auguste de, J. Cambessédes, and Adrien de Jussieu. 1829–32. *Flora Brasiliae Meridionalis*, V1–3 (Apud A. Belin: Paris).

Schiebinger, Londa. 2003. 'Jeanne Baret: the first woman to circumnavigate the globe', *Endeavour*, 27: 22–25.

Seba, Albertus. 1734–65. *Das Naturaliekenkabinett* (Janssonio-Waesbergois *et al.*: Amersterdam).

Seth, Catriona. 2012. 'Nobody's children? Enlightenment foundlings, identity and individual rights', *Burgerhart Lectures Dutch-Belgian Society for Eighteenth-Century Studies*, 5: 4–52.

Sheridan, Geraldine. 2009. *Louder than Words: Ways of seeing women workers in eighteenth-century France* (Texas Tech University Press: Lubbock).

Sibbald, Susan, and Frances Paget Hett. 2007. *The Memoirs of Susan Sibbald 1783–1812* (Kessenger Publishing: Whitefish, Montana).

Siembieda, William J. 1996. 'Walls and gates: A Latin perspective', *Landscape Journal*, 15: 113–122.

Smeaton, William A. 1988. 'La Pérouse and the scientific exploration of the Pacific, 1785–1788', *Endeavour*, 12: 34–37.

Snell, Hannah, Mary Lacy, and Mary Anne Talbot. 2008. *The Lady Tars: The Autobiographies of Hannah Snell, Mary Lacy and Mary Anne Talbot* (Fireship Press: Tucson).

Stevenson, Robert L. 1896. 'Lay Morals', *The Works of Robert Louis Stevenson*, V4, (T. & A. Constable: London).

——. 1908. 'The Dangerous Archipelago—atolls at a distance', Part II: 'The Paumotus', *In the South Seas* (Chatto and Windus: London).

Sulloway, Frank J. 1982. 'Darwin and his finches: The evolution of a legend', *Journal of the History of Biology*, 15.1: 1–53.

Taillemite, Étienne. 1977. *Bougainville et ses Compagnons autour du monde: 1766–1769, Journaux de Navigation*, V1–2 (Imprimerie Nationale: Paris).

Taylor, Paul, D. 1998. 'Fossils in folklore', *Geology Today*, 14(4): 142–145.

Tepe, Eric J., Glynis Ridley, and Lynn Bohs. 2012. 'A new species of *Solanum* named for Jeanne Baret, an overlooked contributor to the history of botany', *PhytoKeys*, 8: 37–47.

Théré, Christine. 1999. 'Women and birth control in eighteenth-century France', *Eighteenth-Century Studies*, 32.4: 552–564.

UNODC. 2018. *Global Study on Homicide: Gender-related killing of women and girls* (United National Office on Drugs and Crime: Vienna), www.unodc.org/documents/data-and-analysis/GSH2018/GSH18_Gender-related_killing_of_women_and_girls.pdf.

van Tilburg, Marja. 2006. 'The allure of Tahiti: Gender in late eighteenth-century French texts on the Pacific', *History Australia*, 3: 1–16.

Vermeij, Geerat J. 2010. 'Sound reasons for silence: Why do molluscs not communicate acoustically?', *Biological Journal of the Linnean Society*, 100: 485–493.

Verne, Jules. 1890. 'Souvenirs d'enfance et de jeunesse', Part 2. jv.gilead.org.il/garmt/souvenirs2.html.

Veselka, Vanessa. 2014. 'Green screen: The lack of female road narratives and why it matters', *The American Reader*, 1.4: 9.

Vinson, Auguste. 1874. 'Un compagnon de voyage autour du monde', *Revue des Sociétés Savantes*, 8 (Series 10): 291–297.

Wallis, Helen. ed. 1965. 'Editorial', in *Carteret's Voyage Round the World 1766–1769*, V1 (Hakluyt Society: Cambridge).

Wheelwright, Julie. 1989. *Amazons and Military Maids: Women who dressed like men in pursuit of life, liberty and happiness* (Pandora: London).

Williams, Glyn. 2013. 'Scurvy on the Pacific voyages in the age of Cook', *Journal for Maritime Research*, 15: 37–45.

Williams, Roger L. 2003. *French Botany in the Enlightenment: The ill-fated voyages of La Pérouse and his rescuers*, International Archives on the History of Ideas Series, No. 182 (Kluwer: Dordrecht).